CAMBRIDGE TRACTS IN MATHEMATICS

---

77. *Approaches to the theory of optimization*

J. PONSTEIN
Professor at the University of Groningen

Approaches to the theory of optimization

CAMBRIDGE UNIVERSITY PRESS
CAMBRIDGE
LONDON NEW YORK NEW ROCHELLE
MELBOURNE SYDNEY

PUBLISHED BY THE PRESS SYNDICATE OF THE UNIVERSITY OF CAMBRIDGE
The Pitt Building, Trumpington Street, Cambridge, United Kingdom

CAMBRIDGE UNIVERSITY PRESS
The Edinburgh Building, Cambridge CB2 2RU, UK
40 West 20th Street, New York NY 10011–4211, USA
477 Williamstown Road, Port Melbourne, VIC 3207, Australia
Ruiz de Alarcón 13, 28014 Madrid, Spain
Dock House, The Waterfront, Cape Town 8001, South Africa

http://www.cambridge.org

First published 1980
First paperback edition 2004

A catalogue record for this book is available from the British Library

ISBN 0 521 23155 8 hardback
ISBN 0 521 60491 5 paperback

Contents

Preface

Perhaps the title of this book should have been, somewhat in the style of past times, 'Some approaches towards the theory of optimization, with an emphasis on the topological aspects, ignoring combinatorial problems and almost ignoring combinatorial tools, not going into the algorithmic and numeric problems of effectively finding solutions to problems, yet meant as a contribution to applied and even practical mathematics.'

Even this does not make it sufficiently clear what is going on in the book and what is not. It is worth saying a few words about the omissions. Convex processes are not treated, problems involving more than one objective to be optimized are only touched on lightly. Of game theory, only the simplest model is considered. Since, as far as Lagrangian duality is concerned, any practitioner wants a nonzero multiplier attached to the objective function, theorems where this multiplier is allowed to be zero do not receive any attention. Similarly the reader will not find anything about regularity conditions, which apart from being sufficient are also necessary for a whole class of problems (rather than for a single one). This is because when we try to solve any one problem out of such a class, those conditions may well be too strong for practical purposes. This is not to say that everything in the book is so 'practical': some basic theorems rest on the axiom of choice, for example, and sometimes we are satisfied to establish the equality of an infimum and a supremum rather than that of the corresponding minimum and maximum.

Although in practical situations we can often make do with decision variables which are elements of a Banach space, or even a Euclidean space, part of the theory includes a generalization to locally convex topological vector spaces. This provides us with the important possibility of considering, in Banach spaces, topologies weaker than the strong one.

If we single out for competition the results dealt with in the appendixes, the prize for the most beautiful result should perhaps go to Theorem 6.2.2 on the equality of *inf sup* and *sup inf*, because of its generality and its natural and elegant proof; or perhaps to the general fixed point theorem (6.1.18) because of its combinatorially ingenious proof; or to the theory of conjugate duality (chapter 4) because of its power and symmetry. The reader should judge for himself.

The main text does not contain references to the literature. These, together with comments, have been combined in a separate section entitled 'Comments on the text and related literature'. Internal cross-references are indicated either by two numerals, such as 3.14 referring to section 14 of chapter 3, or by three numerals, such as 3.14.19 referring to an item in 3.14.

The text assumes a basic knowledge of topology as well as functional analysis.

Explicitly, J. W. Nieuwenhuis contributed by generalizing some basic theorems, and implicitly he and W. K. Klein Haneveld contributed through many discussions on all kinds of subjects involved. The number of these discussions is only countable, but their effect was invaluable.

Symbols

Latin

A usually a matrix
B usually a matrix
b usually a right-hand side
C usually a subset of X
cl closure of
$\mathrm{D}h(z,\tau)$ see (4.4.4)
dom effective domain
epi epigraph
f objective function
F bifunction
F^{d} dual bifunction
F^{dd} bidual bifunction
g in $g(x) \leqslant 0$
g_{a} (includes) the active part of g (in $g(x) \leqslant 0$)
G set (feasible region)
h in $h(x) = 0$ or a function
h^* conjugate of h
h^{**} biconjugate of h
$h'(z_{\mathrm{o}}, z)$ one-sided directional derivative
H hyperplane or Hamiltonian
int interior
$K(\gamma)$ cone defined by (3.8.6)
L Lagrangian
l_1, l_∞ see Example 3.2.9
$L(V)$ see Definition 3.3.4
N negative cone or nullspace
N^* dual negative cone
p perturbation function
p^{d} dual perturbation function
p^{dd} bidual perturbation function
P positive cone
P^* dual positive cone
q' usually Fréchet derivative of q
R real axis
ri relative interior
R_n n-dimensional Euclidean space
T set defined in (3.5.3) or point-to-set mapping defined by (6.1.7)
U a set or a neighbourhood
u control variable, or in $u \in U$
V a set or a neighbourhood, or as defined by (3.5.4)
W a set or a neighbourhood
X decision space
x in $x \in X$, decision variable
X^* dual perturbation space
x^* in $x^* \in X^*$, dual perturbation
Y constraint space = perturbation space

y in $y \in Y$, (primal) perturbation
Y^* multiplier space
y^* in $y \in Y^*$, multiplier
Z a (perturbation) space
z in $z \in Z$
Z^* dual of Z
z^* in $z^* \in Z^*$

Greek

α infimum as in (3.5.1)
β supremum as in (3.5.2)
δ indicator function
δ^* support function

Other

∇ gradient
∂ subgradient

1

Approaching optimization by means of examples

1.1 Definition of optimization problem

The subjects we want to consider are centred around the problem of finding the *infimum*, or if it exists, the *minimum*, of a given real-valued function f on a given set G of a given space X. Hence if we let $f: x \in X \mapsto f(x) \in R$ and define α by

$$(1.1.1) \qquad \alpha = \inf\{f(x): x \in G\}, \quad G \subset X,$$

then the underlying problem is that of finding the value of α, or, if it exists, an $x_o \in G$ such that $f(x_o) = \alpha$. We shall refer to this basic problem as an *optimization problem*. Usually X is called the *space of decisions* or *decision space*, and $x \in X$ is called the *decision variable*, although when $x = (\xi_1, \ldots, \xi_n) \in R_n$, the components of x are also termed decision variables, so that this term is slightly ambiguous.

Understandably, in practical situations we want an $x_o \in G$ such that $f(x_o) = \alpha$. Such an x_o is called an *optimal solution*, or simply a *solution*, whereas any x in G is called a *feasible solution*. We say that our optimization problem is *solvable* if at least one optimal solution exists, and that it is *feasible* if at least one feasible solution exists.

The function f is the *objective function*. As stated before, its values are real numbers. In a more general type of optimization problem, however, where $f(x)$ is not a real number but, say, an s-dimensional vector, it is not surprising that f is then termed the *multi-objective function*. A fair question now would be how to define an optimization problem with many objectives. In order to achieve this, one must generalize the idea of infimum. We shall come back to this more general question in 7.5 but it will only be touched on. See also Example 1.2.9.

The set G is the *constraint set* or *feasible region*. A very special example is where $G = X$, in which case we speak of an *unconstrained* problem. This is a natural terminology, since the larger G is, the fewer restrictions there are. As a second example, let functions $g_i: X \to R$, $i = 1,\ldots,p$ and $h_j: X \to R$, $j = 1,\ldots,q$ be given and let G be defined by

(1.1.2) $$G = \{x: g_i(x) \leqslant 0,\ i = 1,\ldots,p, \quad h_j(x) = 0,\ j = 1,\ldots,q\}.$$

In particular, if the g_i are absent, we arrive at the abstract formulation of many classical optimization problems in mechanics and other areas, problems which are often solved by the method of Lagrange multipliers. The subject of this book is nothing more than the generalization of ideas and results involving these multipliers!

Each separate condition $g_i(x) \leqslant 0$ or $h_j(x) = 0$ is called a *constraint*, or more specifically an *inequality constraint* or an *equality constraint*. Defining $g: X \to R_p$ and $h: X \to R_q$ by $g(x) = (g_1(x),\ldots,g_p(x))$ and $h(x) = (h_1(x),\ldots,h_q(x))$, and writing $g_i(x) \leqslant 0$, $i = 1,\ldots,p$, as $g(x) \leqslant 0$, which is now an inequality relation between vectors, we can further write

(1.1.3) $$G = \{x: g(x) \leqslant 0, \quad h(x) = 0\}.$$

Somewhat inconsistently, $g(x) \leqslant 0$ and $h(x) = 0$ are also often called *constraints*, although strictly speaking each of these consists of a number of constraints.

Instead of (1.1.3) we frequently come across the following form of G,

(1.1.4) $$G = \{x: g(x) \leqslant 0, \quad h(x) = 0, \quad x \in C\},$$

where C is some given set of X. The main reason for slipping in another set is not so much that it would be impossible to define G completely in terms of inequality and equality constraints alone, but that we want to treat the constraints $g(x) \leqslant 0$ and $h(x) = 0$ differently from the constraint $x \in C$. In fact we shall not be too concerned about the nature of C, although general conditions (such as being convex) may be imposed on it.

Instead of an infimum, we may want to find a *supremum*. Obviously, all we have said so far can be carried over to supremum problems, if only because of the following trivial relation

(1.1.5) $$\sup\{f(x): x \in G\} = -\inf\{-f(x): x \in G\}.$$

In practical situations X is often a Euclidean space or more generally a Banach space. Many of the theorems that follow, however, hold if X is a locally convex topological vector space. These more general spaces can be used fruitfully in a number of cases where the norm topology of a Banach space is replaced by a weaker one. The latter may be necessary in order that the dual space is a convenient one.

The variety of optimization problems in areas such as mathematics, engineering, economics and statistics is enormous. In order that the reader gets at least some idea of this variety, and to illustrate the meaning of the definitions given so far, we list a few of them in the next section.

1.2 Examples of optimization problems

1.2.1 **Example.** Given a closed set in a Euclidean space, find the least distance from the origin to that set. For example, if $x \in R_n$, A is an $m \times n$ matrix, $b \in R_m$ and $|Ax-b|$ is the Euclidean norm of $Ax-b$, find

$$\inf\{|Ax-b|: x \in R_n\}.$$

Then the said Euclidean space is R_m, and the closed set is $\{Ax-b: x \in R_n\}$. This is just the simple least squares approximation. If we replace the Euclidean spaces by Banach spaces this example becomes what is known as the *minimum norm problem*. This is further discussed in Example 3.14.19.

1.2.2 **Example.** The *Chebyshev approximation problem* is concerned with approximating some function q by means of a given finite set of functions by minimizing the maximum of the absolute value of the difference of $q(t)$ and a linear combination of the given functions. As an example let $q: [-1, +1] \to R$ and let the given functions be $1, t, \ldots, t^n$, $t \in [-1, +1]$. Then the problem is to find

$$\inf\{f(x): x \in R_{n+1}\},$$

where $x = (\xi_0, \ldots, \xi_n)$ and

$$f(x) = \max\left\{\left|q(t) - \sum_{i=0}^{n} \xi_i t^i\right|: -1 \leqslant t \leqslant +1\right\}.$$

Both this example and the previous one are examples of unconstrained optimization problems. The Chebyshev approximation problem is worked out in 7.2.

1.2.3 **Example.** Let $a_1, \ldots, a_n$ be given positive numbers whose sum is 1. Find

$$\sup\{\xi_1^{a_1} \ldots \xi_n^{a_n}: a_1 \xi_1 + \ldots + a_n \xi_n = 1 \xi_i \geqslant 0, \quad i = 1, \ldots, n\}.$$

As will be shown in 7.1, the optimal solution is given by $\xi_i = 1$ for all i, which implies that

$$\xi_1^{a_1} \ldots \xi_n^{a_n} \leqslant a_1 + \ldots + a_n \xi_n \quad \text{for all} \quad \xi_i \geqslant 0, \quad i = 1, \ldots, n.$$

1.2.4 **Example.** Suppose a horizontal, rigid, homogeneous steel plate, which has the shape of a polygon with four vertices, is supported at these vertices. At a given point T of the plate a vertical force P is applied to it. This force is directed downwards. Obviously, P results in forces ξ_i acting on the supports, $i = 1, \ldots, 4$, and $P = \Sigma \xi_i$. The ith support remains rigid as long as $\xi_i \leqslant F_i$, where F_i is known, but suddenly breaks down as soon as $\xi_i > F_i$. Find the maximum P such that no support will break down. If we introduce a coordinate system in the plane of the plate such that $T = (0, 0)$ and if we let (a_i, b_i) be the ith vertex, and then observe the conditions for all forces to be in equilibrium, the problem becomes that of finding

$$\sup\{\Sigma \xi_i: \Sigma a_i \xi_i = 0, \quad \Sigma b_i \xi_i = 0, \quad \xi_i \leqslant F_i, \quad i = 1, \ldots, 4\}.$$

Notice that this problem has a very special form: everything is linear, moreover there are only a finite number of constraints, and $x = (\xi_1, \ldots, \xi_4)$ is finite dimensional. Problems of this type are termed *linear programming problems*, and are treated in 3.13.

1.2.5 **Example.** Linear programming problems, as defined in the previous example, arise very often in economics. Consider the following so-called *production planning problem*. Let $x = (\xi_1, \ldots, \xi_n)$ represent the amounts ξ_j of n goods that must be produced, let $b = (\beta_1, \ldots, \beta_m)$ represent the maximal amounts β_i of m raw materials that may be used for producing the goods, and let the production of one unit of the jth good require the amounts α_{ij} of raw material i, $i = 1, \ldots, m$. The price at which the jth good can be sold on the market is γ_j per unit. Find an optimal production plan, that is find x, such that $\Sigma\gamma_j\xi_j$ is as large as possible; hence find

$$\sup\{\Sigma\gamma_j\xi_j: \Sigma\alpha_{ij}\xi_j \leqslant \beta_i, \quad i = 1, \ldots, m, \quad \xi_j \geqslant 0, \quad j = 1, \ldots, n\}.$$

or, putting $c = (\gamma_1, \ldots, \gamma_n)$ and letting α_{ij} be the ijth element of a matrix A, find

$$\sup\{cx: Ax \leqslant, x \geqslant 0\},$$

where, just as in (1.1.3), we have used inequality relations between vectors. Notice the very special kind of constraint $x \geqslant 0$, or $\xi_j \leqslant 0$, $j = 1, \ldots, n$. These are called *nonnegativity constraints*.

1.2.6 **Example.** So far our examples have involved *finite*-dimensional decision spaces only, which would give the wrong impression that we could restrict ourselves to Euclidean spaces. Here is a much more general example. Let t represent time and let a certain physical system (a rocket to one of the earth's satellites) be described adequately if its so-called *state* $x(t)$ is known at all relevant times. The state is thought of as a finite-dimensional vector, and might consist of position and velocity of the system. The system can be controlled by the *control* u, which like x is a function of t, and $u(t)$ is also taken to be a finite-dimensional vector. The meaning of the control is that if u is specified the state $x(t)$ is also fixed, or, in other words, that $x: t \mapsto x(t)$ can be solved uniquely from a certain set of equations $g(x, u) = 0$. The control might include the position of steering mechanisms, the rate of fuel consumption, and the like. It is quite probable that limitations are set to u, as obviously follows from the examples suggested. Let us summarize those by $u \in G$. Finally, $f(x, u) \in R$ represents some measure of performance of the system, say the time required to reach a target, or the total fuel consumption. At any rate we assume that the system performs best if $f(x, u)$ is minimal. Then the problem is to find

$$\inf\{f(x, u): g(x, u) = 0, \quad u \in G\}.$$

To be more specific, assume that $f(x, u)$ is some integral over a fixed period of time $[0, T]$, ruling out the possibility that $f(x, u)$ is the time to reach a target, and assume that x can be solved from an integral equation; then given T, G, x_0, ϕ and ψ the problem is to find

$$\inf\left\{\int_0^T \phi(x(t), u(t), t)\,\mathrm{d}t: x(t) = x_0 + \int_0^T \psi(x(t), u(t), t)\,\mathrm{d}t, u \in G\right\}.$$

This is a typical *fixed time optimal control problem.* A simpler problem is obtained if the integral equation, which is equivalent to a differential equation with initial conditions, is replaced by difference equations. A more complex problem arises if this differential equation is replaced by a *partial* differential equation.

The question of what the decision space should be can be answered in two ways. Either it is the space of the controls u, in which case the equation by which x can be solved in terms of u should not be considered a constraint, but merely a *defining equation*; or it is the product of the space of the controls and that of the states, in which case $g(x, u) = 0$ becomes a *constraint.* If x can be easily solved from this equation then the former approach looks better, but if solving for x is a difficult matter then it may be wise to follow the latter approach, where (x, u) becomes the decision variable, *not u.* A simple control problem is solved in 7.3.

1.2.7 **Example.** Consider two persons, playing the following, perhaps not very exciting, *game*. Player i must select an element x_i from a given set G_i of, say, a Euclidean space X_i, $i = 1, 2$. They must do so independently from each other, that is to say the one player does not know what the other is selecting. Further, a function $\phi: X_1 \times X_2 \to R$ is given. After x_1 and x_2 have been chosen, player 1 must pay the amount $\phi(x_1, x_2)$ to player 2. End of the game!

Now what is the optimization problem here? Obviously player 1 wants to minimize $\phi(x_1, x_2)$, but at the same time player 2 wants to maximize this amount. So there are two conflicting objectives. Even if we forget about player 2 and consider only player 1's desire to minimize $\phi(x_1, x_2)$, there is the problem that he does not know anything about x_2. To resolve the difficulty, let us assume that player 1 adopts the following reasoning: if I select x_1, then the maximum loss I might incur is $\sup_{G_2} \phi(x_1, x_2)$, so let me select x_1 such that I minimize this supremum. This leads to the problem of finding

$$\inf_{G_1} \sup_{G_2} \phi(x_1, x_2),$$

where $\sup_{G_2} \phi(x_1, x_2)$ is the objective function. A similar reasoning in player 2's mind leads to the problem of finding

$$\sup_{G_2} \inf_{G_1} \phi(x_1, x_2),$$

where now $\inf_{G_1} \phi(x_1, x_2)$ is the objective function.

Hence we have managed to create *two* optimization problems. An interesting question in the *theory of games* is under what conditions a pair $(x_1^\circ, x_2^\circ) \in G_1 \times G_2$ exists such that

$$\sup_{G_2} \inf_{G_1} \phi(x_1, x_2) = \phi(x_1^\circ, x_2^\circ) = \inf_{G_1} \sup_{G_2} \phi(x_1, x_2)$$

or equivalently

$$\phi(x_1^\circ, x_2) \leqslant \phi(x_1^\circ, x_2^\circ) \leqslant \phi(x_1, x_2^\circ)$$

$$\text{for all } x_1 \in G_1 \text{ and all } x_2 \in G_2.$$

Because of the last relation such a pair is called a *saddle-point of* ϕ (with respect to G_1 and G_2).

Quite apart from the theory of games (a subject we shall not pursue very deeply) the existence of saddle-points will be one of the vital facts we shall be interested in. Then the function ϕ will be replaced by what will be termed the *Lagrangian function*.

Finally we remark that the theory of games encompasses a little more than just this very simple game!

1.2.8 **Example.** Let us now turn to a *stochastic* optimization problem. It is an example of *inventory control*, well known in economics. During a given number of periods a certain item has to be produced so as to satisfy a demand for it. The demand presents itself at the *end* of each period and is stochastic. If the demand at the end of any period turns out to be less than what is available at that moment then inventory costs are incurred, and if the demand is more than what is available, then shortage costs are incurred. Paying these costs does not mean that the surplus or the shortage disappears; on the contrary they accumulate. Apart from inventory costs and shortage costs we are faced with production costs. Assume for simplicity that only two periods are considered, that all costs are linear and the same for both periods, that c is the cost of production per unit, h is the inventory cost per unit and u is the shortage cost per unit. Further assume that at the beginning of period 1 there is neither a positive inventory, nor a positive shortage. Finally, and this is an important assumption, the production in the second period may depend on the actual demand in the first period, so that the problem is one with '*recourse*'.

The problem is how to produce in order that the *expected* total cost is minimized.

Let S be the set of possible demands, e.g. $S = [0, 10]$ or $[0, +\infty)$, etc. and let μ be the probability measure of the demand distribution. Define the function $(\cdot)^+$ by

$$r^+ = r \quad \text{if } r \geqslant 0, \quad r^+ = 0 \quad \text{if } r < 0, \quad r \in R.$$

Then the problem is to find

$$\inf\Big\{cx_1+\int_S [cx_2(s)+h(x_1-s)^+ + u(s-x_1)^+]\,\mu(\mathrm{d}s)$$
$$+\int_S\int_S [h(x_1+x_2(s)-s-s')^+ + u(s+s'-x_1-x_2(s))^+]\,\mu(\mathrm{d}s')\,\mu(\mathrm{d}s):$$
$$x_1 \geqslant 0,\quad x_2(s)\geqslant 0 \text{ for (almost) all } s\Big\}.$$

The decision variable is $x=(x_1, x_2)$ where $x_1 \in R$ but x_2 is a function on S to R. Hence X is an infinite-dimensional space, unless the given probability distribution is discrete and finite.

Usually this sort of problem is solved by *dynamic programming*, but in 7.4 we shall view the present problem from the standpoint taken in this book, that is from the standpoint of *mathematical programming* (see 2.9).

1.2.9 **Example.** This example, too, is stochastic in nature and arises in statistical decision theory. Let the possible outcomes of a random experiment form the set S. S is known, but the underlying probability distribution is not, although it is known that it is one out of M completely known distributions, which we indicate by $D_1, \ldots, D_M$. For simplicity we let $S=\{s_1, \ldots, s_N\}$, for some N; hence S is finite. With D indicating the unknown distribution, we let p_{mn} be the probability that the outcome is s_n when $D=D_m$, $n=1,\ldots,N$, $m=1,\ldots,M$. Clearly, all p_{mn} are known. Further let Q, $1\leqslant Q<M$ be given and consider two hypotheses H and K. H is the *null-hypothesis*: $D=D_m$ for some m satisfying $1\leqslant m\leqslant Q$; whereas K is the *alternative hypothesis*: $D=D_m$ for some m satisfying $Q+1\leqslant m\leqslant M$. Statisticians speak of making *an error of the first kind* if H is rejected, when it is in fact true, and of making *an error of the second kind* if H is accepted, when it is in fact false. They introduce a *testfunction* x from S to, say, the set $\{0, 1\}$, to serve the following purpose. Suppose a single sample is drawn from S. If $x(s_n)=0$ then H is accepted, if $x(s_n)=1$ then H is rejected. In case $D=D_m$ *and* $1\leqslant m\leqslant Q$ then $\Sigma_{n-1}^{N} p_{mn}x(s_n)$ is the probability of making an error of the first kind. This probability is kept low by simply putting an upper bound, say 0.05, to it. And in case $D=D_m$ but

$Q+1 \leqslant m \leqslant M$ then $1-\Sigma_{n=1}^{N} p_{mn} x(s_n)$ is the probability of making an error of the second kind. Instead of putting bounds on this probability as well, it will be minimized. This would lead, however, to *multi-objective optimization* unless we take $Q = M-1$. With this further restriction we are thus led to the problem of finding

$$\sup\left\{\sum_{n=1}^{N} p_{Mn} x(s_n): \sum_{n=1}^{N} p_{mn} x(s_n) \leqslant 0.05, m = 1, \ldots, M-1; \right.$$
$$\left. x(s_n) = 0 \text{ or } 1, \quad n = 1, \ldots, N\right\}.$$

Since all relations are linear, except for the constraint $x(s_n) = 0$ or 1, this problem is an example of what is known as *integer (linear) programming*; 'integer' because the decision variables are required to be integer valued. Such problems are out of the scope of this book. If however, we replace x by a function from S to $[0, 1]$, then the result is linear programming, which is one of the simplest applications of what is going to follow. Other variations are obtained by taking S and/or D infinite. Then the decision space X becomes infinite-dimensional or the number of constraints becomes infinite, or both. Yet linearity remains. In these cases one speaks of *(one-sided) infinite linear programming problems*, at least if $x: S \to [0, 1]$.

2

An intuitive approach to mathematical programming

2.1 Introduction

In this chapter we begin by considering the following two problems.

(A) Let us assume that we are given Euclidean spaces $X = R_n$ and $Y = R_m$, a mapping $f\colon X \to R$, a mapping $g\colon X \to Y$, and a subset C of X. The problem is to find

$$\alpha = \inf\{f(x)\colon g(x) \leqslant 0, x \in C\}, \tag{2.1.1}$$

where $g(x) \leqslant 0$ means that each component of $g(x)$ is nonpositive. Clearly (2.1.1) is just (1.1.4) with $X = R_n$ and h left out.

(B) Assume now that we are given a Euclidean space $X = R_n$, a mapping $f\colon X \to R$ and a subset G of X. The problem is to find

$$\alpha = \inf\{f(x)\colon x \in G\}. \tag{2.1.2}$$

We want to know what conclusions can be drawn if these problems are *perturbed* in a certain way, by studying the sensitivity of the infimum with respect to the perturbations.

2.2 Perturbation of the two problems

(A) First we consider (2.1.1). We want to know what happens if we replace $g(x) \leqslant 0$ by $g(x) - y \leqslant 0$ or, equivalently, by $g(x) \leqslant y$, where y is some fixed element of $Y = R_m$. Notice that, as mentioned after (1.1.4), we treat the two contraints, namely $g(x) \leqslant 0$ and $x \in C$, differently. The former is perturbed, but the latter is not. Define $p\colon Y \to R$ by

$$p(y) = \inf\{f(x)\colon g(x) \leqslant y, x \in C\}, \quad y \in Y. \tag{2.2.1}$$

This is the *perturbation function*, which will play an essential part in what follows, at least in a generalized version. Obviously $p(0) = \alpha$, $p(y) \leqslant \alpha$ if $y \geqslant 0$, and $p(y) \geqslant \alpha$ if $y \leqslant 0$. If $y \neq 0$ in the last two cases, then in general, we will have that $p(y) < \alpha$ or that $p(y) > \alpha$. For this reason we add a positive or negative 'cost' for the very act of perturbation of the problem. Let y_0^* be some linear functional on R_m, so that y_0^* is an element of the *dual* R_m^* of R_m. We could, of course, identify R_m^* with R_m, but we want to distinguish the two because of generalizations we have in mind, generalizations where such an identification would not be possible. Take y_0^* nonnegative, add $y_0^* y$ to $p(y)$, and compare the sum with α. Suppose we could find y_0^* (the 'cost per unit perturbation') such that no perturbation whatsoever would be 'profitable', that is such that

$$(2.2.2) \quad p(y)+y_0^* y \geqslant p(0) = \alpha \quad \text{for all } y \in Y = R_m, \quad y_0^* \geqslant 0,$$

or equivalently

$$(2.2.3) \qquad \inf\{p(y)+y_0^* y \colon y \in Y\} = \alpha, \quad y_0^* \geqslant 0.$$

If we eliminate $p(y)$ by means of (2.2.1) and interchange the order of the two resulting inf operations, it easily follows that

$$(2.2.4) \qquad \inf\{f(x)+y_0^* g(x) \colon x \in C\} = \alpha,$$

for y_0^* is nonnegative. The perturbation has completely disappeared, but y_0^* is still there. Ignoring the whole area of perturbing problems, we could have simply asked whether a nonnegative y_0^* exists such that (2.2.4) holds. Later on we shall see that this is so if, for example, f and g are convex functions, C is a convex set, and certain regularity conditions are satisfied.

Notice that in (2.2.4) $g(x) \leqslant 0$ is no longer a constraint; we got rid of it by introducing the *multiplier* y_0^*, and arrived at a *global condition*, at least with respect to C. If $C = X = R_n$ then we would have obtained an unconstrained optimization problem.

If (2.2.4) holds for some $y_0^* \geqslant 0$ and if x_o is an optimal solution, then $f(x_o)+y_0^* g(x_o) \geqslant f(x_o)$, hence $y_0^* g(x_o) \geqslant 0$. But we also have that $y_0^* g(x_o) \leqslant 0$ since $g(x_o) \leqslant 0$, hence

$$(2.2.5) \qquad y_0^* g(x_o) = 0,$$

which is a very useful relationship. It is closely related to what is known as the *complementary slackness* condition of linear programming (see. 3.13) and the *transversality* condition of optimal control. It says that if the ith component of $g(x_o)$ is negative, the ith component of y_0^* must be zero. Intuitively this is obvious, because we could delete the constraint $g_i(x) \leqslant 0$ if $g_i(x_o) < 0$ and still have that x_o is optimal. Deleting a constraint, however, amounts to equating its multiplier to zero. Conversely, if the ith component of y_0^* is positive, then $g_i(x_o) = 0$; that is the constraint $g_i(x) \leqslant 0$ is *active* or *binding* at x_o.

(A′) If we replace the inequality constraint $g(x) \leqslant 0$ by the equality constraint $h(x) = 0$, and put

(2.2.6) $$p(y) = \inf\{f(x)\colon h(x) = y, \quad x \in C\},$$

then we cannot, of course, claim that $p(y) \leqslant \alpha$ if $y \geqslant 0$, or that $p(y) \geqslant \alpha$ if $y \leqslant 0$, since then we have neither enlarged nor diminished the feasible region, but shifted it. For this reason we must drop the requirement that y_0^* should be nonnegative, but this is the only change we have to make in the reasoning under (A) above, although (2.2.5) then tells us nothing new, because $g(x_o) = 0$ if x_o is optimal.

(A″) But let us return to case (A), and add a few (rather restrictive) conditions: $C = R_n$; f and g are convex functions, hence for any λ, $0 \leqslant \lambda \leqslant 1$ we have that

$$f(\lambda x + (1-\lambda)x') \leqslant \lambda f(x) + (1-\lambda) f(x')$$

and $$g(\lambda x + (1-\lambda)x') \leqslant \lambda g(x) + (1-\lambda) g(x')$$

for all x and all x'; f and g are differentiable; and there exists an optimal solution x_o. Letting $\lambda > 0$ tend to zero and using the assumed convexity and differentiability we obtain

$$f(x) - f(x_o) \geqslant [f(x_o + \lambda(x - x_o)) - f(x_o)]/\lambda$$

and

(2.2.7) $$f(x) - f(x_o) \geqslant \nabla f(x_o)(x - x_o) \quad \text{for all } x,$$

and similarly

(2.2.8) $$g(x_o) - g(x_o) \geqslant \nabla g(x_o)(x - x_o) \quad \text{for all } x.$$

Moreover, if (2.2.4) holds, then $\nabla(f(x)+y_o^* g(x)) = 0$ if $x = x_o$. Postmultiplying this latter equation by $x-x_o$ and applying (2.2.7) we get

$$f(x)-f(x_o) \geqslant -y_o^* \nabla g(x_o)(x-x_o) \quad \text{for all } x. \tag{2.2.9}$$

If $g(x) \leqslant 0$ we have that $-y_o^*(g(x)-g(x_o)) \geqslant 0$, because of (2.2.5). Postmultiplying again by $x-x_o$, we obtain by (2.2.8) that

$$-y_o^* \nabla g(x_o)(x-x_o) \geqslant 0 \quad \text{for all } x \text{ such that } \quad g(x) \leqslant 0. \tag{2.2.10}$$

Setting $x_o^* = -y_o^* \nabla g(x_o)$ we get rid of g as well as differentials:

$$\begin{aligned} & f(x)-f(x_o) \geqslant x_o^*(x-x_o) \quad \text{for all } x, \text{ and} \\ & x_o^*(x-x_o) \geqslant 0 \quad \text{for all } \textit{feasible } x. \end{aligned} \tag{2.2.11}$$

Having arrived at this result, we would like to know whether it was really necessary to make such strong conditions. As we shall see later, we can indeed drop a number of these conditions; in particular it is not required to assume differentiability and, what is perhaps more important for the present discussion, the constraint set need not be specified by means of inequalities, but may be any convex set G. On the other hand certain regularity conditions must be imposed in order for (2.2.11) to hold for some x_o^*.

Comparing (2.2.11) with (2.2.4) which, if $C = X = R_n$ and if an optimal solution x_o exists, is equivalent to

$$f(x)-f(x_o) \geqslant -y_o^* g(x) \quad \text{for all } x, \tag{2.2.12}$$

we see that the part played by y_o^* is taken over by x_o^*. Since $x_o^* \in R_n^*$, the dual of R_n, we could try to arrive at (2.2.11) in a more direct way by taking $Y = R_n$ instead of taking $Y = R_m$, which brings us back to the idea of introducing an appropriate perturbation. Since (2.2.11) tells us that the feasible region need not have a special form, let us change the scene and replace the problem of finding

$$\alpha = \inf\{f(x): g(x) \leqslant 0, \quad x \in C\}$$

by that of finding $\alpha = \inf\{f(x): x \in G\}$; see (2.1.2).

(B) Let α now be defined by (2.1.2) and let us try to introduce a perturbation y, which must be an element of $R_n = X$. The right choice turns out to be

(2.2.13) $\quad p(y) = \inf\{f(x): x+y \in G\}, \quad y \in Y = R_n = X.$

Again we add to $p(y)$ a cost of perturbation, namely $x_0^* y$, where $x_0^* \in R_n^*$, and compare $p(y)+x_0^* y$ with α. Again assuming that an x_0^* exists such that no perturbation whatsoever will pay, we get instead of (2.2.3) the following equation:

$$\inf\{p(y)+x_0^* y: y \in Y\} = \alpha$$

or
$$\inf_{x,y}\{f(x)+x_0^* y: y \in Y,\ x+y \in G\} = \alpha$$

or
$$\inf_{x,y}\{f(x)+x_0^*(y-x): y \in Y, y \in G\} = \alpha$$

or, if $G \neq \varnothing$,

(2.2.14) $\quad \inf_x\{f(x)-x_0^* x: x \in X\}+\inf_y\{x_0^* y: y \in G\} = \alpha,$

which replaces (2.2.4). The reader might remark that (2.2.4) is no more *global* than (2.1.2), our starting point. He should notice, however, that the first term in the left hand side of (2.2.14) is an infimum over the entire space X, and that the objective in the second term is a very special function. Moreover in the first term, G does not occur, and in the second term f does not occur.

Now let us see what happens if there is an optimal solution x_0. Then (2.2.14) becomes

(2.2.15) $\quad f(x)-f(x_0) \geqslant x_0^*(x-y) \quad$ for all $x \in X$ and all $y \in G$.

Taking $y = x_0$ we get the first inequality of (2.2.11) and taking $x = x_0$ we get the second, where 'feasible' now means that $x \in G$. Hence, we have indeed found a direct argument to arrive at (2.2.11). In fact, we have found more, namely (2.2.14), which does not require the existence of an optimal solution x_0.

2.3 Introducing perturbations abstractly; bifunctions

Once again comparing (2.2.4) and (2.2.14) where the existence of an optimal x_0 is not assumed, or (2.2.12) and (2.2.15) where the existence of x_0 is assumed, we see that the method of perturbation of a problem greatly influences the results (none of them have been proved yet!). In principle, any type of perturbation will do. As a third example of perturbing a problem let us consider the Chebyshev approximation problem of Example 1.2.2. Notice that this problem led to an unconstrained optimization problem. We perturb it by perturbing the function q. Hence let y be a function on $[-1, +1]$ to R, and perturb the problem by replacing $q(t)$ by $q(t)+y(t)$. Clearly Y, which is the space of y's, is a function space. This space is not yet defined completely but that is not our concern at this moment (see 7.2 for more details in this respect). The unperturbed Chebyshev approximation problem leads to the problem of finding

$$\inf\{f(x): x \in X = R_{n+1}\} \tag{2.3.1}$$

with f defined as in Example 1.2.2, whereas the perturbed problem takes the following form: find

$$\inf\{F(x, y): x \in X\}, \quad y \in Y. \tag{2.3.2}$$

This formulation leads us immediately to an abstract generalization of the perturbation of optimization problems. Ignoring how F was derived in the Chebyshev approximation problem, all we have to do is perturb the problem by introducing a function F. In other words, let a decision space X *and* a linear space Y (the space of *perturbations*) be given, as well as a function $F: X \times Y \to R$, and consider the (perturbed) problem of finding the infimum of (2.3.2). Call the problem of finding

$$\inf\{F(x, 0): x \in X\} \tag{2.3.3}$$

the unperturbed problem.

In any practical situation where, of course, the unperturbed problem is given, one should select the function F in such a way that the problem of finding (2.3.3) is identical with the given one. Following Rockafellar, we call F the *bifunction* of the problem.

We have, however, created a difficulty by requiring that (2.3.2) should be an unconstrained problem. How do we find Fs corresponding to the problems posed in 2.1? It seems that the only way out is to let F take *infinite* values. The difficulty is now resolved, for in case A we put

$$\begin{aligned} F(x, y) &= f(x) \quad \text{if } g(x) \leqslant y \text{ and } x \in C \\ &= +\infty \quad \text{otherwise} \end{aligned} \tag{2.3.4}$$

and in case B we take

(2.3.5)
$$F(x, y) = f(x) \quad \text{if} \quad x+y \in G$$
$$= +\infty \quad \text{otherwise.}$$

Admittedly infinite function values are not very attractive from a numerical point of view, but they are extremely useful for developing a general theory of duality.

2.3.6 **Definition.** Given a space X and a linear space Y, F is a *bifunction* if F: $X \times Y \rightarrow [-\infty, +\infty]$. Given a bifunction F, the corresponding *perturbation function* is defined by

$$p(y) = \inf\{F(x, y): x \in X\}, \quad y \in Y.$$

Notice that $p(y)$, too, is an element of the extended real line.

Let us again compare $\alpha = \inf\{F(x, 0): x \in X\}$ with $p(y)+y_0^*y$. In order to get proper results we must require that Y is not only a linear but also a topological space, and that y_0^* is not only linear but also continuous. Hence we assume that Y is a *topological vector space*, and that y_0^* is a *linear continuous functional* on Y, hence y_0^* is an element of the dual Y^* of Y with respect to the topology of Y. Reasoning as before we obtain the following result, or rather conjecture since we have not yet given a proof,

(2.3.7)
$$\inf_{x, y}\{F(x, y)+y_0^*y: x \in X, y \in Y\} = \alpha.$$

Specifying F as in (2.3.4) or (2.3.5) we may derive (2.2.4) or (2.2.14) from (2.3.7).

So far, three spaces play a part: the *decision space* X, the *perturbation space* Y, and its dual Y^*, which we could call the *multiplier space*. Later on we shall see that the dual of X will also play a part. Then we must, of course, assume that X as well as Y is a topological vector space.

In many practical applications Y (and also X) will be a Banach space. With little extra effort, however, we can let Y be a locally convex topological vector space. This at the same times gives us the important opportunity to ignore the natural topology of a Banach space, namely its norm topology or strong topology, and to consider a weaker, or *the* weak, topology instead. The reason why this is so important is that the dual of a Banach space with respect to its strong topology may turn out to be an undesirable space. We shall give an example of this later on in 3.2. This is not to say that using weaker topologies is always a good idea, for sets with nonempty interior with respect to the strong topology may cease to have such an interior if we change from the strong topology to a weaker one. And the existence of a nonempty interior is required in certain separation theorems, which we need in certain duality results!

2.4 Subdifferentiability

Perhaps the (mathematical) reader does not appreciate the introduction of the 'cost of perturbation'. Let us then try to follow another line of thought assuming, however, that (in the case of problem 2.1.1) f and g are convex functions and that C is a convex set. Then it turns out that p is also a convex function. If, in addition, p is differentiable (which even in a simple case like linear programming is in general not so), then analogous to (2.2.7) we find

$$(2.4.1) \qquad p(y)-p(0) \geqslant \nabla p(0)\, y \quad \text{for all} \quad y \in R_m.$$

Since $p(y) \leqslant p(0) = \alpha$ for all $y \geqslant 0$, it follows from this that $\nabla p(0) \leqslant 0$ (take $y \geqslant 0$ and take any one component of y arbitrarily large). Hence $y_0^* = -\nabla p(0) \geqslant 0$ and $p(y)-p(0) \geqslant -y_0^* y$, which is (2.2.2) again.

This reasoning not only may suit the more formal reader but also suggests an important generalization of differentiability. For suppose that p is not differentiable, but that nevertheless a y_0^* exists such that $p(y)-p(0) \geqslant -y_0^* y$ for all y; then we can consider this as a generalization of differentiability. Such a y_0^* is called a *subgradient* of p at 0. It is not always unique.

2.5 The Lagrangian

In 1.1 we made the remark that classical optimization problems of the form $\inf\{f(x): h(x) = 0\}$ are often solved by means of the method of Lagrange multipliers. This means that the Lagrangian function $L(x, y^*) = f(x) + y^* h(x)$ is formed, the derivative of $L(x, y^*)$ with respect to x is set equal to zero, and x and y^* are solved from the resulting equations and $h(x) = 0$.

We shall use the term Lagrangian function, or simply Lagrangian, for a more general expression. Let $F: X \times Y \to [-\infty, +\infty]$ be given, with Y a topological vector space.

2.5.1 **Definition.** The *Lagrangian* $L: X \times Y^* \to [-\infty, +\infty]$ (with respect to F) is defined by

$$L(x, y^*) = \inf\{F(x, y) + y^*y : y \in Y\}, \quad x \in X,\ y^* \in Y^*.$$

If we define F by

$$(2.5.2) \qquad \begin{aligned} F(x, y) &= f(x) && \text{if} \quad h(x) = y \\ &= +\infty && \text{otherwise}, \quad x \in R_n,\ y \in R_m, \end{aligned}$$

then we quickly find from this definition that $L(x, y^*) = f(x) + y^*h(x)$, which is what we had before.

Now let us take

$$(2.5.3) \qquad \begin{aligned} F(x, y) &= f(x) && \text{if} \quad g(x) \leqslant y \quad \text{and} \quad x \in C \\ &= +\infty && \text{otherwise}, \quad x \in R_n,\ y \in R_m; \end{aligned}$$

then we get

$$(2.5.4) \quad \begin{aligned} L(x, y^*) &= f(x) + y^*g(x) && \text{if} \quad x \in C \quad \text{and} \quad y^* \geqslant 0 \\ &= -\infty && \text{if} \quad x \in C \quad \text{and} \quad y^* \ngeqslant 0 \\ &= +\infty && \text{if} \quad x \notin C, \end{aligned}$$

which shows what happens if we replace equality constraints by inequality constraints and at the same time add the constraint $x \in C$.

As a third example, let

$$(2.5.5) \qquad \begin{aligned} F(x, y) &= f(x) && \text{if} \quad x + y \in G \\ &= +\infty && \text{otherwise}, \quad x, y \in R_n, \end{aligned}$$

then

$$(2.5.6) \qquad L(x, x^*) = f(x) - x^*x + \inf\{x^*y : y \in G\},$$

where, as before, we have put x^* instead of y^*, since $Y = X$.

Notice that, whereas with (2.5.2) L is linear in y^*, and with (2.5.3)

L is linear in y^* for $x \in C$ and $y^* \geq 0$, with (2.5.5) L in general is not linear in x^*. In all cases, however, L is concave in y^* or x^*.

Let us now assume that an optimal solution x_o exists, as well as a y_0^* such that (2.3.7) holds, and hence that

$$\text{(2.5.7)} \qquad \inf_{x,y}\{F(x,y)+y_0^* y: x \in X, y \in Y\} = F(x_o, 0).$$

Then

$$\begin{aligned} L(x_o, y_0^*) - L(x, y_0^*) &= \inf\{F(x_o, y)+y_0^* y: y \in Y\} \\ &\quad -\inf\{F(x,y)+y_0^* y: y \in Y\} \leq F(x_o, 0) \\ &\quad -\inf_{x,y}\{F(x,y)+y_0^* y: x \in X, y \in Y\} = 0, \end{aligned}$$

and

$$\begin{aligned} L(x_o, y^*) - L(x_o, y_0^*) &= \inf\{F(x_o, y)+y^* y: y \in Y\} \\ &\quad -\inf\{F(x_o,y)+y_0^* y: y \in Y\} \leq F(x_o, 0) \\ &\quad -\inf_{x,y}\{F(x,y)+y_0^* y: x \in X, y \in Y\} = 0, \end{aligned}$$

so that

$$\text{(2.5.8)} \quad L(x_o, y^*) \leq L(x_o, y_0^*) \leq L(x, y_0^*), \quad \text{for all } \; x \in X \; \text{ and all } \; y^* \in Y^*.$$

This means that (x_o, y_0^*) is a *saddle-point* of the Lagrangian. Moreover, (2.5.7) implies that

$$F(x_o, 0) = \inf_x L(x, y_0^*) \leq L(x_o, y_0^*) = \inf_y (F(x_o, y)+y_0^* y) \leq F(x_o, 0)$$

so that

$$F(x_o, 0) = L(x_o, y_0^*) = \inf_x L(x, y_0^*).$$

In other words, if (2.5.7) holds then x_o is a minimum of the Lagrangian with $y^* = y_0^*$ fixed.

We still have to keep in mind that we have not proved that (2.5.7) holds; the proof requires nontrivial assumptions. If we reverse the reasoning however, in certain cases we can easily prove the following statement, without any additional assumptions.

Let F and L be given by (2.5.3) and (2.5.4) (where we tacitly assumed that $C \neq \varnothing$ and that $f(x)$ and $g(x)$ are finite if $x \in C$) and suppose that (x_o, y_0^*) is a saddle-point of L. Then x_o is an optimal solution of the problem of finding $\inf\{f(x): g(x) \leq 0, x \in C\}$, $y_0^* \geq 0$ and $\inf\{f(x)+y_0^* g(x): x \in C\} = f(x_o)$. To see that this is true, first notice that $x_o \in C$, for if not then $L(x, y_0^*) = +\infty$ for all x, and hence $C = \emptyset$. Also $y_0^* \geqslant 0$, for if not then $L(x_0, y^*) = -\infty$ for all y^*; in particular $y^* = 0$. This means that we can simplify (2.5.8) as follows.

(2.5.9) $$f(x_0)+y^*g(x_0) \leqslant f(x_0)+y_0^*g(x_0) \leqslant f(x)+y_0^*g(x) \quad \text{for all} \quad x \in C \quad \text{and all} \quad y^* \geqslant 0.$$

If we take $y^* = 0$, it follows from this that $y_0^*g(x_0) \geqslant 0$ and, if we take any one component of y^* arbitrarily large, that $g(x_0) \leqslant 0$, so that x_0 is feasible and $y_0^*g(x_0) \leqslant 0$, and hence that $y_0^*g(x_0) = 0$. Hence $f(x_0) = \inf\{f(x)+y_0^*g(x): x \in C\}$, and the optimality of x_0 easily follows.

A similar statement can be made if we let

$$L(x, y^*) = f(x)+y^*h(x).$$

But consider now the third example, namely (2.5.5) and (2.5.6), and suppose (x_0, x_0^*) is a saddle-point of L. Then (2.5.8) implies that

(2.5.10) $$f(x_0)-x^*x_0+\inf\{x^*y: y \in G\} \leqslant f(x_0)-x_0^*x_0 +\inf\{x_0^*y: y \in G\} \leqslant f(x)-x_0^*+\inf\{x_0^*y: y \in G\}.$$

The second inequality here immediately leads to the first one of (2.2.11), and taking $x^* = 0$ in (2.5.10) we have that

$$f(x_0) \leqslant f(x_0)-x_0^*x_0+\inf\{x_0^*y: y \in G\} \leqslant f(x_0)-x_0^*x_0+x_0^*y \quad \text{if} \quad y \in G,$$

and from this we find the second inequality of (2.2.11). Taking $x^* = 0$ again we also have that

$$f(x_0) \leqslant f(x)-x_0^*x+\inf\{x_0^*y: y \in G\} \leqslant f(x) \quad \text{if} \quad x \in G,$$

so that $f(x_0) = \inf\{f(x): x \in G\}$. But we cannot show that x_0 is feasible! In fact examples exist where x_0 is infeasible.

2.5.11 **Counterexample.** Let $x \in R$, $f(x) = x$ and $G = \{x: x > 0\}$. Then it is easily verified that $(x_0, x_0^*) = (0, 1)$ is a saddle-point of L, but no optimal solution exists at all, hence x_0 cannot be optimal.

It is immediately clear what is wrong with this example: G is not closed. This suggests that if any bifunction is given and if (x_0, y_0^*) is a saddle-point of the corresponding Lagrangian, then we shall need some sort of condition about closures to be able to conclude that x_o is optimal. It is perhaps tempting to try and deal with (2.5.11) by defining $F(x, y) = x$ if $x > y$, and $F(x, y) = +\infty$ otherwise, which is similar to (2.5.3), but this is of no help, as can be verified without much effort.

Notice that, even although this counterexample is somewhat abnormal, an x_o^* still exists such that (2.2.14) holds (with $\alpha = 0$).

2.6 Duality

We can write the saddle-point condition (2.5.8) more compactly as

$$(2.6.1) \qquad \inf_X L(x, y_o^*) = L(x_o, y_o^*) = \sup_{Y^*} L(x_o, y^*)$$

which implies that

$$(2.6.2) \qquad \sup_{Y^*} \inf_X L(x, y^*) = \inf_X \sup_{Y^*} L(x, y^*),$$

since we always have that $\sup_{Y^*} \inf_X L(x, y^*) \leqslant \inf_X \sup_{Y^*} L(x, y^*)$, and since

$$\begin{aligned} \inf_X \sup_{Y^*} L(x, y^*) &\leqslant \sup_{Y^*} L(x_o, y^*) \\ &= \inf_X L(x, y_o^*) \leqslant \sup_{Y^*} \inf_X L(x, y^*). \end{aligned}$$

For (2.6.2) to be true it is sufficient that (2.3.7) holds, i.e. that for some y_o^* we have that

$$\inf_X F(x, 0) = \inf_{X,Y} (F(x, y) + y_o^* y) = \inf_X L(x, y_o^*);$$

the existence of x_o is not necessary. For we always have that $L(x, y^*) \leqslant F(x, 0)$ so that

$$\begin{aligned} \inf_X \sup_{Y^*} L(x, y^*) \leqslant \inf_X F(x, 0) &= \inf_X L(x, y_o^*) \\ &\leqslant \sup_{Y^*} \inf_X L(x, y^*). \end{aligned}$$

Expressing $\sup_{F^*} L(x, y^*)$ in terms of F we find that it is equal to

$$\sup_{Y^*} \inf_Y (F(x, y) + y^* y).$$

If we assume for a moment that here as in (2.6.2) we may interchange sup and inf, the latter expression becomes $\inf_Y \sup_{Y^*} (F(x, y) + y^* y)$, which is equal to $F(x, 0)$ if $F(x, y) > -\infty$ for all y, and equal to $-\infty$ if $F(x, y) = -\infty$ for some y. Notice that x is fixed here. Hence under certain conditions we may conclude that $\sup_{Y^*} L(x, y^*) = F(x, 0)$, which is the objective of the unperturbed problem, and that finding $\inf_X \sup_{Y^*} L(x, y^*)$ amounts to finding $\inf_X F(x, 0)$ which is the unperturbed problem itself.

It is inviting now, in view of the symmetry of (2.6.2), to define

$$\phi(y^*) = \inf_X L(x, y^*) \tag{2.6.3}$$

as the objective function of some *other* optimization problem, and to consider the problem of finding

$$\beta = \sup_{Y^*} \phi(y^*) = \sup_{Y^*} \inf_X L(c, y^*). \tag{2.6.4}$$

Doing so we obtain the following theorem.

2.6.5 **Theorem.** If, for some y_o^*,

$$\inf_{X, Y} (F(x, y) + y_o^* y) = \alpha = \inf_X F(x, 0),$$

then y_o^* is an optimal solution of the problem of finding β, defined by (2.6.4), and moreover $\alpha = \beta$.

This result is of fundamental importance for what lies ahead, and to stress this we introduce the following terminology.

The problem of finding $p(y) = \inf_X F(x, y)$ is termed the (perturbed) *primal problem*, whereas the problem of finding $\beta = \sup_{Y^*} \phi(y^*)$ is termed the (unperturbed) *dual problem*, regardless of whether $\alpha = p(0)$ is equal to β or not, and if so, whether the supremum is attained at some y_o^* or not.

A number of natural questions arises:

(*a*) Under what conditions is $\alpha = \beta$ and is the supremum a maximum? This is only restating a question we posed before.

(*b*) Under what conditions can we also be certain that an optimal solution x_o exists of the primal problem?

Or if we weaken our requirements:

(*c*) Under what conditions is $\alpha = \beta$?

(*d*) If $\alpha = \beta$ and if optimal solutions x_o and y_o^* exist, is it meaningful to solve the dual problem instead of the primal, and how do we recover x_o if y_o^* has been computed? (We must be careful here because although we may have that $L(x_o, y_o^*) = \inf_x L(x, y_o^*)$, there may be nonoptimal x' with $L(x', y_o^*) = \inf_x L(x, y_o^*)$. So when minimizing $L(x, y_o^*)$ we must pick the right solution. Several people circumvent the difficulty here by assuming that the minimum of $L(x, y_o)$ is unique, which is true if $L(\cdot, y_o)$ is strictly convex.)

Answering these questions at least partly will be the main content of the chapters to follow.

2.7 Perturbation of the dual problem

The symmetry of (2.6.2) led us to the introduction of the dual problem, but clearly it is an unperturbed problem. If we want more symmetry we should try to introduce a certain perturbation into the dual problem. In order to achieve this let us write β in full, in terms of F, that is

$$(2.7.1) \qquad \beta = \sup_{Y^*} \inf_{X,Y} (F(x,y) + y^*y)$$

and let us recall that three spaces were involved, namely X, Y and Y^*. In order to obtain a certain symmetry we assume that X, as well as Y, is a topological vector space, introduce the term x^*x where $x \in X^*$, the dual of X), and define the *dual perturbation function* p^d by

$$(2.7.2) \qquad p^d(x^*) = \sup_{Y^*} \inf_{X,Y} (F(x,y) - x^*x + y^*y)$$

and the *dual bifunction* F^d by

$$(2.7.3) \qquad F^d(y^*, x^*) = \inf_{X,Y} (F(x,y) - x^*x + y^*y).$$

Clearly, the pair (X, Y) is now replaced by the pair (Y^*, X^*); in the dual problem Y^* is the decision space, and X^* is the perturbation space, hence we should indeed write $F^d(y^*, x^*)$, and *not* $F^d(x^*, y^*)$. (Instead of *dual* perturbation function and *dual* bifunction some authors speak of *adjoint* perturbation function and *adjoint* bifunction.)

The minus sign in $-x^*x$ is not essential, but leads to a desirable result if we apply our definitions to (finite-dimensional) *linear programming*. Then

$$(2.7.4) \qquad \begin{aligned} F(x,y) &= cx && \text{if } \; y + Ax \geqslant b \; \text{ and } \; x \geqslant 0 \\ &= +\infty && \text{otherwise} \end{aligned}$$

($x \in R_n$, $c \in R_n^*$, $y \in R_m$, $b \in R_m$, A an $m \times n$ matrix), so that

(2.7.5) $$p(y) = \inf\{cx: y+Ax \geqslant b, x \geqslant 0\}.$$

Then

$$\inf_{X,Y}(F(x,y)-x^*x+y^*y)$$
$$= \inf_{x,y}\{cx-x^*x+y^*y: y+Ax \geqslant b, x \geqslant 0\}.$$

If $y^* \ngeqslant 0$ then this is equal to $-\infty$, and if $y^* \geqslant 0$ then this is equal to

$$\inf_x\{(c-x^*)x+y^*(b-Ax): x \geqslant 0\},$$

which is equal to y^*b if $x^*+y^*A \leqslant c$ and $-\infty$ if $x^*+y^*A \nleqslant c$. Hence, by (2.7.3) we have that

(2.7.6) $$F^{\mathrm{d}}(y^*,x^*) = y^*b \quad \text{if} \quad x^*+y^*A \leqslant c \quad \text{and} \quad y^* \geqslant 0$$
$$= -\infty \quad \text{otherwise,}$$

and

(2.7.7) $$p^{\mathrm{d}}(x^*) = \sup_{y^*}\{y^*b: x^*+y^*A \leqslant c, y^* \geqslant 0\}.$$

Comparing these two equations with (2.7.4) and (2.7.5) we observe certain differences: inf is replaced by sup, $+\infty$ by $-\infty$, x by y^*, y by x^*, c by b, b by c, a $\geqslant$ inequality by a $\leqslant$ inequality, and transposing everything we see that really A is replaced by its transpose.

If we had taken $+x^*x$ instead of $-x^*x$, then the results would not have become as symmetric as they are now.

This example also shows that in certain cases we may expect perfect duality, since the primal is the dual of the dual. The simplest way to verify this is perhaps to write the dual problem as an infimum problem, to transpose everything, and to apply the dualization anew, keeping in mind that the dual of a Euclidean space can be identified with that space itself.

Returning to the general case, it trivially follows from the definitions that $\beta = p^{\mathrm{d}}(0)$, so that $\alpha = \beta$ is equivalent to $p(0) = p^{\mathrm{d}}(0)$. Further it follows that $p^{\mathrm{d}}(0) = \sup_{Y^*}\inf_Y(p(y)+y^*y)$, and if we once more assume that we may interchange sup and inf here, this is equal to $\inf_Y\sup_{Y^*}((p(y)+y^*y)$, which is equal to $p(0)$ if $p(y) > -\infty$ everywhere, and equal to $-\infty$ if $p(y) = -\infty$ somewhere. Hence if the interchange of sup and inf is allowed, and if $p(y) > -\infty$ everywhere, then $p(0) = p^{\mathrm{d}}(0)$. This shows how important it is to know under what conditions sup inf = inf sup, and that something else interferes, namely whether or not $p(y) > -\infty$ for all y. We leave this for chapter 4.

Finally we ought to point out that in general our relationships are not completely symmetric, as they are in a simple case like linear programming. For in order to construct the dual of the dual we must replace (Y^*, X^*) by (X^{**}, Y^{**}), so that the dual of the dual can only be equated to the primal if we may identify X^{**} with X and Y^{**} with Y. When this is so we say that X and X^{**} are *compatible* and similarly for Y and Y^{**}. In chapter 3 we shall see that sometimes if we work with the strong topology of Banach spaces X cannot be identified with X^{**} or that Y cannot be identified with Y^{**}, but that we can make everything compatible if we consider weaker topologies (with the risk, as remarked before, that certain sets may no longer possess a nonempty interior).

2.8 Global conditions versus local conditions

Let us return to one of the special problems we considered in 2.2, namely the problem of finding

(2.8.1) $$p(y) = \inf\{f(x) \colon g(x) \leqslant y, x \in C\}.$$

The (unperturbed) dual of this problem is to find

(2.8.2) $$\sup_{Y^*} \phi(y^*) = \sup\{\inf\{f(x) + y^* g(x) \colon x \in C\} \colon y^* \geqslant 0\},$$

and, as we have seen, the existence of an optimal solution y_0^* of the dual problem is tantamount to the existence of a y_0^* such that

(2.8.3) $$\inf\{f(x) + y_0^* g(x) \colon x \in C\} = \alpha,$$

which is (2.2.4) and which we have interpreted as a *global* condition.

Let us assume that the primal problem, too, has an optimal solution, say x_0, f and g are differentiable and moreover C is a convex set (i.e. $\lambda x + (1-\lambda) x' \in C$ whenever $x, x' \in C$ and $0 \leqslant \lambda \leqslant 1$). Then if $x \in C$ so is $x_0 + \lambda(x - x_0)$, if $0 < \lambda \leqslant 1$, so that (2.8.3) and $y_0^* g(x_0) = 0$ imply that

$$f(x_0 + \lambda(x - x_0)) + y_0^* g(x_0 + \lambda(x - x_0)) \geqslant f(x_0) + y_0^* g(x_0),$$

which leads to

(2.8.4) $$\nabla_x (f(x_0) + y_0^* g(x_0))(x - x_0) \geqslant 0 \quad \text{for all} \quad x \in C.$$

Although for (2.8.3) to hold one must assume something like the convexity of f, g and C, it turns out that convexity assumptions for f and g are not always necessary for (2.8.4) to hold. This is not very surprising, since this relationship can be considered to be derived from the problem of finding

$$(2.8.5) \quad \alpha' = \inf\{\nabla f(x_o)(x-x_o): \\ g(x_o)+\nabla g(x_o)(x-x_o) \leqslant 0, x \in C\},$$

at least if it could be shown that $\alpha' = 0$. And if C is convex, everything is convex here, because except for the constraint $x \in C$ everything is linear. We have, of course, done nothing other than linearize the given problem, by using only values of f and g in the vicinity of x_o, i.e. by taking a *local* point of view.

Not only is the convexity of f and g irrelevant. To a certain extent this can also be said of the convexity of C, since we need only take $x' = x_o$ in the inclusion $\lambda x + (1-\lambda)x' \in C$. Hence what really matters is the differentiability. Moreover many problems relate to functions that are not convex, but are differentiable. For this reason (2.8.4) is very important. It is, for example, related to the *Pontryagin minimum principle* of optimal control, and this is why several authors speak of (2.8.4) as the *minimum principle* (or pre-minimum principle) relating to the given optimization problem.

That we are taking a *local* point of view becomes even more clear if we assume that x_o is an interior point of C, for example because C is open. Then we simply get that

$$(2.8.6) \qquad \nabla_x(f(x_o)+y_o^* g(x_o)) = 0,$$

which is a *local optimality condition*. This and the conditions $y_o^* g(x_o) = 0$, $g(x_o) \leqslant 0$ and $x_o \in C$ are together usually termed the *Kuhn–Tucker conditions*.

2.9 Mathematical programming versus dynamic programming

We can now define what we mean by *mathematical programming*. It is that branch of pure and applied mathematics where one investigates the problem of finding $\inf_X F(x, 0)$ with the aim of obtaining results involving the Lagrangian function $L(x, y^*) = \inf_Y (F(x, y) + y^*y)$. This means that the *multiplier* y^* must play a crucial part. But since this multiplier is coupled to the *perturbation* y via the bilinear form y^*y, perturbations also play a vital part. In a sense perturbations play a more important part than do multipliers, because when examining a problem one will usually first answer the question 'where and how should I perturb my problem?' In many cases the answer to this question will completely define the perturbation space Y, and hence Y^* which is the multiplier space, so that multipliers are *derived* from perturbations; although in other cases there will be some freedom to choose Y (in particular its topology) and one will use this freedom to arrive at an acceptable Y^*. On the other hand perturbations do not usually occur in final results, whereas multipliers do; moreover these multipliers serve as the variables of the dual problem, if this can be defined.

The results from the theory of mathematical programming roughly fall into two categories: global results making possible the definition of a dual problem next to the given primal problem, and local results which can often be interpreted as duality theorems of a linearized version of the given problem and its dual.

At one time there was a tendency to regard a problem as a mathematical programming problem only when the spaces involved were finite-dimensional, and perhaps there are still people who use the term in the restricted sense. Terms like *linear programming*, *nonlinear programming*, *integer programming* and *convex programming* should be immediately clear now, except that *quite a few* people are using these terms in the restricted sense, in particular the terms linear programming and integer programming.

In order to show more clearly the content of the term mathematical programming, let us contrast it with that of *dynamic programming*. Whereas mathematical programming may lead to algorithms for solving optimization problems as well as to all kinds of duality results, dynamic programming immediately takes the form of an algorithm. In fact it is a very general method for solving optimization problems, but because it is so general only its scheme can be indicated. Suppose, for instance, that we are given a minimization problem involving the real variables $x_1, \ldots, x_N$. Further suppose that we are able to set up the following recursive procedure (where the sign $:=$ should be read as an instruction to assign to the left-hand side the value of the right-hand side; so if before the execution of $i := i+1$ we had $i = 4$ then after it we would have $i = 5$)

1. $f_0(s_0) := 0$;
2. $i := 0$;
3. $i := i+1$;
4. compute $f_i(s_i) := \min_{x_i} [a_i(s_i, x_i) + f_{i-1}(s_{i-1})]$ with $s_{i-1} := g_i(s_i, x_i)$ for all 'relevant' values of s_i;
5. if $i = N$ stop, if not go back to 3;

with the following stipulations: $f_i(s_i)$ must be the minimum of a suitably defined 'subproblem' involving only the variables $x_1, \ldots, x_i$; certain parameters in this subproblem are represented by s_i; and $f_N(s_N)$ is the minimum of the given problem.

The idea is to relate a problem with variables $x_1, \ldots, x_i$ to what we hope are less complicated problems with variables $x_1, \ldots, x_{i-1}$. And the problem is how to define the functions f_i, a_i and g_i, as well as s_i which is the *state variable of dynamic programming*. Once everything is properly defined we go through the algorithm above to obtain $f_N(s_N)$ as well as the optimal solution for $(x_1, \ldots, x_N)$.

It is difficult to say in general terms what s_i should be, and $f_i(s_i)$, etc. Only when a problem is completely specified can these questions be answered.

We shall, however, not go into the details of dynamic programming. All we shall do is indicate a relationship between the $f_i(s_i)$ and a certain linear programming problem, whose optimal dual variables are exactly these $f_i(s_i)$; see 7.6.

Finally we should remark that the scheme given above is not the only possible dynamic programming scheme. Another important class of dynamic programming algorithms is where step 4 is replaced by

4. compute $f_i := \min_x [a_i(x) + f_{i-x}]$ $(x = 1, 2, \ldots, i)$

and where s_i is deleted altogether.

3

A global approach by bifunctions

3.1 Convexity and the positive cone

Since a general definition of convex *functions* relies on that of what we shall call the positive cone, which in turn relies on the definition of convex *sets*, we shall treat first convex sets, then the positive cone, and finally convex functions.

3.1.1 **Definition.** Let Z be a linear space and D be a subset of Z. Then D is a *convex set* of Z if $\lambda z+(1-\lambda)z' \in D$, whenever $z, z' \in D$ and $0 \leqslant \lambda \leqslant 1$. Given $z_i \in Z$, $\lambda_i \geqslant 0$, $i = 1,\ldots,n$ such that $\Sigma\lambda_i = 1$, $\Sigma\lambda_i z_i$ is a *convex combination* of $z_1,\ldots,z_n$. The *convex hull* of any $D \subset Z$ is the set of all convex combinations of pairs of elements of D, or equivalently, it is the intersection of all convex sets of Z containing D.

This definition enables us to introduce the idea of a positive cone. The reason why we want such a cone is so that we can generalize the constraint $g(x) \leqslant 0$, where $g(x) \in R_m$, to a similar constraint when $g(x)$ is an element of some topological space. Since if $g(x) \in R_m$, $g(x) \leqslant 0$ means that $g(x) \in -K$, where $K = \{y: y \geqslant 0\}$, and since K is a convex cone in this case, we are led to the following definitions.

3.1.2 **Definition.** A set K of a linear space Z is called a *cone with apex at the origin*, or simply a *cone* (as we shall avoid considering cones with apexes elsewhere), if $0 \in K$, and if $\lambda z \in K$, whenever $z \in K$ and $\lambda > 0$.

3.1.3 **Definition.** Let Z be a topological vector space, and let one *convex cone* P be singled out, $P \subset Z$, such that $P \neq Z$; then we say that P is the *positive cone* (of Z) and that Z is a *space with* *positive cone* P. Writing $\operatorname{int} P$ for the interior of P, we define the following relations:

$$\begin{aligned} z \geqslant z' &\quad \text{if} \quad z - z' \in P, \\ z > z' &\quad \text{if} \quad z - z' \in \operatorname{int} P, \\ z \leqslant z' &\quad \text{if} \quad z' \geqslant z, \\ z < z' &\quad \text{if} \quad z' > z. \end{aligned}$$

The condition that $p \neq Z$ serves to avoid the undesirable relation $0 > 0$.

Since constraints often have the form $g(x) \leqslant 0$, it is convenient to introduce the negative cone as well.

3.1.4 **Definition.** If P is the positive cone, then $N = -P$ is the *negative cone*.

3.1.5 **Examples**

(A) Let Z be any topological vector space of infinite sequences $z = (\zeta_1, \zeta_2, \ldots)$. Then the usual definition is

$$P = \{z\colon z = (\zeta_1, \zeta_2, \ldots), \quad \zeta_i \geqslant 0, \text{ all } i\}.$$

If $Z = l_\infty$, the space of all absolutely bounded sequences with $|z| = \sup_i |\zeta_i|$ as norm, then $\operatorname{int} P = \{z\colon \zeta_i > \epsilon$, for some $\epsilon > 0$ and all $i\}$. If, on the other hand, $Z = l_1$, the space of all absolutely convergent sequences with $\Sigma_i |\zeta_i|$ as norm, then $\operatorname{int} P = \varnothing$. These examples show that we should be careful and not define the $>$ relation component-wise in all cases.

(B) If Z is a topological vector space of functions $z\colon [0, 1] \to R_m$, then a useful definition is $P = \{z\colon z(t) \geqslant 0$ for (almost) all $t \in [0, 1]\}$.

(C) A somewhat pathological but acceptable example is

$$Z = R_2 \quad \text{and} \quad P = \{z\colon z = (\{\zeta_1, \zeta_2), \zeta_1 \geqslant 0\}$$

or even $P = \{z\colon z = (\zeta_1, \zeta_2), \zeta_1 > 0 \text{ or } z = 0\}$.

(D) An extreme case is obtained if we let $P = \{0\}$. Then $z \geqslant 0$ really means that $z = 0$, which means that we can write equality constraints as 'inequality' constraints.

In chapter 2 we not only came across the vector inequality $g(x) \leqslant 0$, but also $y^* \geqslant 0$. Since if $g(x)$ is in some topological vector space Y, the multiplier y^* of the constraint $g(x) \leqslant 0$ will be in Y^*, we have to introduce a positive cone in Y^* as well as in Y. These two cones, however, cannot be selected independently from each other, as is exemplified by the Euclidean case where $y^* \geqslant 0$ just means that $y^*y \geqslant 0$ for all $y \geqslant 0$, $y \in R_m$, $y^* \in R_m^*$. This at the same time gives us a hint for the correct generalization.

3.1.6 **Definition.** If Z is a topological vector space with positive cone P then the positive cone P^* of Z^*, the dual of Z, is defined by $P^* = \{z^*: z^*z \geqslant 0 \text{ for all } z \in P\}$. P^* is also called the *dual positive cone*.

A related definition in the literature is that of the polar. The *polar* of P is $\{z^*: z^*z \leqslant 0 \text{ for all } z \in P\}$, which is not P^* but the *dual negative cone* $N^* = -P^*$. The term 'polar of P' is also used, however, for the set $\{z^*: |z^*z| \leqslant 1 \text{ for all } z \in P\}$.

We postpone the statement of some theorems regarding positive cones to a later section, since the proof of one of them requires a separation argument. So let us return to convexity.

3.1.7 **Definition.** Let $h: X \to Z$ be a function on a linear space X to a topological vector space Z with positive cone P, and let C be a convex set of X. Then h is a *convex function* (on C) if

$$\lambda h(x) + (1-\lambda)\, h(x') - h(\lambda x + (1-\lambda)\, x') \in P$$

or, equivalently, if

$$h(\lambda x + (1-\lambda)\, x') \leqslant \lambda h(x) + (1-\lambda)\, h(x'),$$

whenever $x, x', \in C$ and $0 \leqslant \lambda \leqslant 1$.

And h is a *concave function* if $-h$ is a convex function.

We have, however, considered functions whose range is not a linear space but the extended real line, namely the bifunction, the perturbation function, and the Lagrangian. No problems arise if in the last definition we replace Z by $[-\infty, +\infty]$ and take $P = [0, +\infty]$, except when there exist x and x' such that $h(x) = +\infty$ and $h(x') = -\infty$. We must, therefore, amend the definition with the clause 'unless $h(x) = +\infty$ (or $-\infty$) and $h(x') = -\infty$ (or $+\infty$)'. A more elegant definition, however, is based on the notion of epigraph.

3.1.8 **Definition.** The *epigraph*, epi h, of a function

$$h: X \to [-\infty, +\infty]$$

is the set epi $h = \{(x, \mu): \mu \in R, \mu \geqslant h(x), x \in X\}$.

Notice that μ must be a real number, and may not take the value $+\infty$.

3.1.9 **Definition.** A function $h: X \to [-\infty, +\infty]$ is a *convex function*, if epi h is a convex set. And again h is a *concave function* if $-h$ is a convex function.

The two definitions can be nicely related to each other if we introduce the notion of effective domain.

3.1.10 **Definition.** The *effective domain*, dom h, of a function $h: X \to [-\infty, +\infty]$ is the set dom $h = \{x: h(x) < +\infty\}$.

Hence if $h: C \to R$, where C is a subset of X, and we define $h(x) = +\infty$ if $x \notin C$, then $C =$ dom h. If the extended function is convex according to Definition 3.1.9, then necessarily C is a convex set and the inequalities of Definition 3.1.7 hold, as can easily be verified.

3.2 Topological considerations

We have already remarked, at the end of 2.3, that it is convenient to require that the spaces we need should be locally convex topological vector spaces. Their definition is as follows.

3.2.1 **Definition.** Z is a *locally convex topological vector space* if Z is a linear space as well as a topological space, possessing a base $\mathscr{U}$ of neighbourhoods of the origin, which satisfy the following requirements.

(1) For all $U \in \mathscr{U}$ and all $V \in \mathscr{U}$ there exists a $W \in \mathscr{U}$ such that $W \subset U \cap V$.

(2) For all $U \in \mathscr{U}$ and all $\alpha \neq 0$, $\alpha U \in \mathscr{U}$.

(3) Any element of $\mathscr{U}$ is a convex set.

(4) For all $U \in \mathscr{U}$ and all $z \in U$, $\lambda z \in U$ if $|\lambda| \leqslant 1$; that is to say, any element of $\mathscr{U}$ is *balanced*, or *circled*.

(5) For all $U \in \mathscr{U}$ and all $z \in Z$, there exists a $\lambda > 0$ such that

$z \in \mu U$ for all μ such that $|\mu| \geqslant \lambda$; which says that any element of $\mathcal{U}$ is *absorbent*.

It is easy to show that any Banach space is a locally convex topological vector space with respect to its norm topology.

Given any Banach space we could, of course, ignore its norm topology and consider another one instead. A topology τ is said to be *weaker* (or *coarser*) than a topology τ' if any τ-open set is τ'-open. Then τ' is called *stronger* (or *finer*) than τ. A space with a relatively strong topology may have an awkward dual with respect to that topology, but may be endowed with a positive cone with a nonempty interior, again with respect to that topology. On the other hand the dual of a space with a relatively weak topology can be prescribed to a certain extent, but may itself not possess a positive cone with a nonempty interior. With this in mind we shall review some types of topology.

Let us first consider the weak topology of a dual pair. In this case two linear spaces, Z and Z^*, are given. The asterisk here does not mean that Z^* is the dual of Z at the moment, but that it will be with respect to the topology we are going to define. It would seem that given Z we could prescribe Z^* completely arbitrarily, but this is not true. The pair (Z, Z^*) must be selected in such a way that there exists a bilinear function $Q: Z \times Z^* \to \mathrm{R}$ satisfying the following conditions.

(3.2.2) For all $z \in Z$, $z \neq 0$, there exists $z_0^* \in Z^*$ such that $Q(z, z_0^*) \neq 0$.
For all $z^* \in Z^*$, $z^* \neq 0$, there exists $z_0 \in Z$ such that $Q(z_0, z^*) \neq 0$.

One says that z_0^* *distinguishes points* of Z and that z_0 *distinguishes points* of Z^*. Instead of $Q(z, z^*)$ we simply write z^*z.

3.2.3 **Definition.** The pair of linear spaces (Z, Z^*) is called a *dual pair* (with respect to Q) if there exists a bilinear function $Q: Z \times Z^* \to R$ satisfying (3.2.2).

We introduce the following topology in Z, called the $\sigma(Z, Z^*)$-topology, or simply the σ-topology.

3.2.4 **Definition.** The base of neighbourhoods of the origin of the $\sigma(Z, Z^*)$-topology is the family of all sets

$$U(A^*) = \{z\colon |z^*z| \leqslant 1 \quad \text{for all } z^* \in A^*\},$$

where A^* is a *finite* subset of Z^*.

3.2.5 **Theorem.** If (Z, Z^*) is a dual pair, then Z is a locally convex topological vector space with respect to the $\sigma(Z, Z^*)$-topology.

Proof. Requirements 4 and 5 of Definition 3.2.1 follow trivially from the definitions, and requirement 3 follows from the linearity of z^*z. As for requirements 1 and 2, let $U = \{z\colon |z^*z| \leqslant 1 \text{ for all } z^* \in A^*\}$ and $V = \{z\colon |z^*z| \leqslant 1 \text{ for all } z^* \in B^*\}$, where both A^* and B^* are finite. Let $W = \{z\colon |z^*z| \leqslant 1 \text{ for all } z^* \in A^* \cup B^*\}$; then $W \subset U \cap V$. And $\alpha U = \{z\colon |z^*z| \leqslant 1 \text{ for all } z^* \in A^*/\alpha\}$, $\alpha \neq 0$.

The next theorem expresses what we really want.

3.2.6 **Theorem.** If (Z, Z^*) is a dual pair then Z^* consists precisely of all σ-continuous linear functionals on Z.

Proof. See Appendix A.

Since the parts played by Z and Z^* are completely symmetric we can also introduce a σ-topology, or more completely a $\sigma(Z^*, Z)$-topology, on Z^* and state theorems similar to Theorems 3.2.5 and 3.2.6 for Z^* instead of Z. We leave this to the reader. The $\sigma(Z, Z^*)$-topology on Z is often called *the weak* topology, and the $\sigma(Z^*, Z)$-topology on Z^* *the weak** topology.

In a sense the σ-topology is the weakest topology we can construct, for given any topology τ on Z, such that for all z^* the linear functional $z \mapsto z^*z$ is τ-continuous, then τ is stronger than σ. The proof of this statement is straightforward and is not given here.

If (Z, Z^*) is a dual pair, stronger topologies on Z can be constructed as follows. For any subset $A^* \subset Z^*$ define $A^{*\circ}$ by

$$A^{*\circ} = \{z\colon \sup_{z^* \in A^*} |z^*z| \leqslant 1\}. \qquad (3.2.7)$$

The set $A^{*\circ}$ is usually called the *polar* of A^* (but note the other meaning given to this term, mentioned in 3.1). Now consider some family $\mathscr{A}^*$ of σ-bounded subsets A^* of Z^*, which means that for each

$A^* \in \mathscr{A}^*$ and each $z \in Z$ we have that $\{\rho: \rho = z^*z \text{ for some } z^* \in A^*\}$ is a bounded set of R.

3.2.8 **Definition.** The base of neighbourhoods of the origin of Z of the *polar topology* (with respect to $\mathscr{A}^*$) is the family of all sets of the form

$$\epsilon \bigcap_{1 \leqslant i \leqslant n} A_i^{*\circ}, \quad \text{where } \epsilon > 0 \text{ and } A_i^* \in \mathscr{A}^*.$$

The strongest polar topology is sometimes called the β-topology. It is obtained by taking for $\mathscr{A}^*$ the set of *all* σ-bounded subsets of Z^*. Unfortunately, the β-dual of Z cannot always be identified with Z^*. The strongest polar topology for which such an identification is possible is the Mackey topology, which is obtained by taking for $\mathscr{A}^*$ the set of all σ-compact subsets of Z^* which satisfy the requirements 3 and 4 of Definition 3.2.1, hence which are both convex and balanced. We forgo the proofs of these statements, for which we refer the reader to textbooks on topology.

In several respects the Mackey topology is the most favourable: it is relatively strong, and the dual with respect to it is a known space, namely Z^*, which may have been prescribed. On the other hand it might not be so easy to find a practical characterization of it.

3.2.9 **Example.** Let Z be the space of all absolutely bounded sequences $z = (\zeta_1, \zeta_2, \ldots)$, and Z^* the space of all absolutely convergent sequences $z^* = (\zeta_1^*, \zeta_2^*, \ldots)$, reserving the symbols l_∞ and l_1, respectively, for when $\sup_i |\zeta_i|$ and $\Sigma_i |\zeta_i|$ are introduced as norms into these spaces (as in Example 3.1.5A). Let $Q(z, z^*) = z^*z = \Sigma_i \zeta_i^* \zeta_i$. With this Q, Z and Z^* form a dual pair. Hence with respect to the σ-topology, as well as with respect to the Mackay topology, Z^* is the dual of Z, and vice versa.

The dual of l_∞, however, which is taken with respect to its norm, cannot be identified with l_1. It can, in fact, be shown (see the example around (3.11.9)) that this dual contains elements z' with the following strange property:

$z'z = 0$ if $z = (\zeta_1, \zeta_2, \ldots)$ and $\zeta_i = 0$ for $i \geqslant i_o$ and
$z'z = 1$ if $z = (1, 1, \ldots)$.

No problems arise if we take the norm-dual of l_1, because it is neatly equal to l_∞.

It can be shown that the β-topology of Z is simply the norm-topology of l_∞, which shows that it is indeed possible that the β-dual of some space cannot be identified with its partner in the dual pair.

We say that locally convex topological vector spaces Z and Z^* are *compatible* if one is the dual of the other and if they satisfy the requirements (3.2.2). The foregoing discussion shows how pairs of compatible spaces may be constructed. Given any locally convex topological vector space Z we can always pair it compatibly with its dual by simply taking $Q(z, z^*) = z^*z$, $z \in Z$, $z^* \in Z^*$. Then one of the requirements (3.2.2) is trivial and the other is a consequence of the Hahn–Banach theorem (see Appendix B).

3.3 Separation theorems

By separation we shall always mean separating two sets V and W of a locally convex topological vector space Z, with dual Z^*, by some hyperplane $H = \{z \colon z^*z = \rho\}$, $z^* \in Z^*$, $\rho \in R$, which in turn means that $z^*z \leqslant \rho$ if $z \in V$ and that $z^*z \geqslant \rho$ if $z \in W$. We shall consider only two types, namely weak separation and strong separation (the adjectives 'weak' and 'strong' have nothing to do with weak or strong topologies). Strong separation is possible if H can be shifted parallel to itself without losing its separating property. If this is not possible we speak of weak separation. A more precise definition is as follows.

3.3.1 **Definition.** Two nonempty sets V and W of a locally convex topological vector space can be *separated strongly* if there exists a nonzero, linear, continuous functional z^* on Z, such that $\sup_{z \in V} z^*z < \inf_{z \in W} z^*z$. And V and W can be *separated weakly* if $\sup_{z \in V} z^*z \leqslant \inf_{z \in W} z^*z$. In either case the hyperplane $H = \{z \colon z^*z = \rho\}$, is called a *separating hyperplane*, or a *hyperplane separating* V and W, if $z^*z \leqslant \rho$ for all $z \in V$ and $z^*z \geqslant \rho$ for all $z \in W$.

Below we list four separation theorems. The *interior* of a set V is indicated by int V.

3.3.2 **Theorem.** Two nonempty closed convex sets V and W of a locally convex topological vector space Z, one of which is compact, can be separated strongly if $V \cap W = \varnothing$.

The compactness cannot be dispensed with here as we see by taking

$$Z = R_2, z = (\zeta_1, \zeta_2), \quad V = \{\zeta: \zeta_1\zeta_2 = 1, \zeta_2 > 0\}$$

and
$$W = \{z: \zeta_1\zeta_2 = -1, \zeta_2 > 0\}.$$

3.3.3 **Theorem.** Two convex sets V and W of a locally convex topological vector space Z can be separated weakly if int $V \neq \varnothing$, $W \neq \varnothing$ and int $V \cap W = \varnothing$.

The condition int $V \neq \varnothing$ cannot be weakened to $V \neq \varnothing$, since in R_2 we cannot separate a convex region W from a line segment V contained in its interior. Nor can we take int $V \cap$ int $W = \varnothing$ instead of int $V \cap W = \varnothing$; for consider the same example again, but with the roles of V and W interchanged. We may, of course, change the conditions to int $V \neq \varnothing$, int $W \neq \varnothing$ and int $V \cap$ int $W = \varnothing$, since if V and int W can be separated, so can V and W.

In many cases int V may be empty, whereas weak separation is still possible. We deal with some of these cases by considering the relative interior.

3.3.4 **Definition.** Let V be a subset of a locally convex topological vector space Z. Define $L(V)$ by $L(V) = \bigcap\{L: L$ is a *closed* linear subspace of Z such that $V \subset v + L$ for some $v \in V\}$. Define the *relative interior* of V (with respect to $L(V)$) by ri $V = \{v: v + U \cap L(V) \subset V$ for some neighbourhood U of the origin of $Z\}$.

Take care to notice that $L(V)$ is a *closed* set. An alternative way of introducing $L(V)$ is by letting the *affine hull* of V be the smallest *linear variety* (which is a translated linear subspace) containing V. Then for each $v \in V$, $v + L(V)$ is the closure of the affine hull of V. Theorem 3.3.3 can now be generalized as follows.

3.3.5 **Theorem.** Two convex sets V and W of a Banach (or at least a Fréchet) space Z can be separated weakly if $\text{ri } V \neq \emptyset$, $W \neq \emptyset$ and if there exists a *closed* linear subspace M of Z such that $L(V)+M = Z$, $L(V) \cap M = \{0\}$ and $(\text{ri } V - W) \cap M = \emptyset$.

Taking $M = \{0\}$ immediately leads to $\text{ri } V = \text{int } V$, and hence to the preceding theorem. A more symmetric result is given in the last of our four separation theorems.

3.3.6 **Theorem.** Two convex sets V and W of a Banach (or at least a Fréchet) space Z can be separated weakly if $\text{ri } V \neq \emptyset$, $\text{ri } W \neq \emptyset$, $\text{ri } V \cap \text{ri } W = \emptyset$, and if $L(V)+L(W)$ is a closed set.

With $\text{ri } V = \text{int } V$ and $\text{ri } W = \text{int } W$ we find something we have already come across when discussing Theorem 3.3.3.

All four theorems depend on the Hahn–Banach theorem, the last two to the open mapping theorem as well, and that is why in the last two theorems we had to assume that Z is a Banach space. Details of the proofs are given in Appendix B.

3.4 Two theorems on the positive cone

Only the second of the following two theorems requires a separation argument.

3.4.1 **Theorem.** Let Z be a locally convex topological vector space with dual Z^* and with positive cone P. Let P^* be the dual positive cone. If $z^* \geqslant 0$, $z^* \neq 0$, and if $z > 0$ (that is if $z^* \in P^*$ and $z \in \text{int } P$, $z^* \neq 0$), then $z^*z > 0$.

Proof. By Definition 3.1.6 $z^*z \geqslant 0$. Suppose that $z^*z = 0$. Since $z \in \text{int } P$, $z + U \subset P$ for some neighbourhood U of the origin of Z. Take any $z' \in Z$. Then, since U is absorbent, it follows from requirement 5 of Definition 3.2.1 that $\epsilon z' \in U$ for some $\epsilon > 0$. Hence $\epsilon z^*z' = z^*(z+\epsilon z') \geqslant 0$, and similarly $\epsilon z^*(-z') \geqslant 0$, so that we have $z^*z' = 0$ for any $z' \in Z$, which implies that $z^* = 0$. But $z^* \neq 0$, hence z^*z cannot be zero.

3.4.2 **Theorem.** Let Z, Z^* and P be as in the preceding theorem. If $z^*z \geqslant 0$ for all $z^* \geqslant 0$, then $z \geqslant 0$ if P is closed.

Proof. Let z_0 be such that $z^*z_o \geqslant 0$ for all $z^* \geqslant 0$, and suppose that $z_0 \ngeqslant 0$, that is $z_0 \notin P$. Then by Theorem 3.3.2 there exists a $z_0^* \in Z^*$ such that $z_0^* z > z_0^* z_0$ for all $z \in P$. Necessarily $z_0^* \geqslant 0$, because otherwise there must exist a $z' \in P$ such that $z_0^* z' < 0$, so that $z_0^*(\lambda z')$ would tend to minus infinity if λ tends to plus infinity. But $\lambda z' \in P$ for all positive λ, so that $z_0^* z_0 < z_0(\lambda z')$ and hence $z_0^* z_0 = -\infty$. Since then $z_0^* \geqslant 0$ it follows from one of the assumptions that $z_0^* z_0 \geqslant 0$. Taking $z = 0$, however, leads to $z_0^* z_0 < 0$ and hence to a contradiction. Thus $z_0 \ngeqslant 0$ cannot be true.

The condition that P is closed cannot, of course, be left out. Notice incidentally, that P^* is always closed with respect to the σ-topology on Z^*, which is obtained by considering the dual pair (Z, Z^*) with respect to the bilinear mapping $Q(z, z^*) = z^*z$.

3.5 A geometric interpretation of inf and sup; normality

Let X, the decision space, be a linear space; let Y, the perturbation space, be a locally convex topological vector space with dual Y^*; and let F: $X \times Y \to [-\infty, +\infty]$ be a bifunction. Then the corresponding primal problem is that of finding

(3.5.1) $$\alpha = \inf_X F(x, 0),$$

and the dual problem is that of finding

(3.5.2) $$\beta = \sup_{Y^*} \inf_{X, Y} (F(x, y) + y^*y).$$

The numbers α and β, or rather the points $(0, \alpha)$ and $(0, \beta)$ with $0 \in Y$, can be given an interesting geometric interpretation if we introduce the following two sets, which we shall be using quite frequently:

(3.5.3) $T = \{(y, \mu): \mu \in R, y = 0 \in Y\}$,

(3.5.4) $V = \{(y, \mu): \mu \in R, y \in Y, \mu \geqslant F(x, y) \text{ for some } x \in X\}$.

Both sets are subsets of $Y \times R$; T is merely the μ-axis and does not involve F, but V does and is convex if F is convex. If $\mu \geqslant F(x, y)$ for some x we have that $\mu \geqslant \inf_X F(x, y) = p(y)$, the value of the perturbation function at y, so V looks very much like epi p. In general, however, the two sets are not identical, as we can see from the following example.

3.5.5 **Counterexample.** Take $X = Y = R$, and $F(x, y) = x$ if $x > y$, and $F(x, y) = +\infty$ otherwise. Then $V = \{(y, \mu): \mu > y)\}$, but $\operatorname{epi} p = \{(y, \mu): \mu \geqslant y\}$.

This example leads us to the following result (where 'cl' means 'closure of').

3.5.6 **Lemma.** Let V and p be as defined above. Then $V \subset \operatorname{epi} p \subset \operatorname{cl} V$, hence $\operatorname{cl} V = \operatorname{cl} \operatorname{epi} p$.

Proof. If $(y, \mu) \in V$, then $\mu \geqslant \inf_X F(x, y) = p(y)$, hence $(y, \mu) \in \operatorname{epi} p$. And if $(y, \mu) \in \operatorname{epi} p$, then $\mu \geqslant p(y) = \inf_X F(x, y)$, hence for all $\epsilon > 0$, $\mu + \epsilon \geqslant F(x', y)$ for some x', which implies that $(y, \mu + \epsilon) \in V$, that is $(y, \mu) \in \operatorname{cl} V$.

This simple lemma is very useful, as are the next two, which are easy to prove.

3.5.7 **Lemma.** Let F and V be as defined. Then

$$\inf_X F(x, 0) = \inf\{\mu: (0, \mu) \in V\}.$$

3.5.8 **Lemma.**

$$\inf_{X,Y} (F(x, y) + y^*y) = \inf\{\mu + y^*y: y \in Y, (y, \mu) \in V\}.$$

Lemma 3.5.7 immediately implies that $(0, \alpha)$ is the '*lowest*' point of $\operatorname{cl}(T \cap V)$. And by Lemma 3.5.8 we can write

$$\begin{aligned}\beta &= \sup_{Y^*} \inf\{\mu + y^*y: y \in Y, (y, \mu) \in V\} \\ &= \sup_{Y^*, \delta \in R} \inf\{\delta: \delta \in R, \mu + y^*y \geqslant \delta, y \in Y, (y, \mu) \in V\}.\end{aligned}$$

The reason why we may set $\mu + y^*y \geqslant \delta$, rather than $\mu + y^*y = \delta$, is that we are taking the supremum over $\delta \in R$. Now

$$H(y^*, \delta) = \{(y, \mu): 1 \cdot \mu + y^*y = \delta\}$$

is a hyperplane in $Y \times R$, and $\mu + y^*y \geqslant \delta$ for all $(y, \mu) \in V$ means that V is on *one* side of it. We might say that V is 'above' the hyperplane, because if $(y, \mu) \in V$ then $(y, \mu + \lambda) \in V$ for all positive λ. Further, it is seen that $(0, \delta)$ is the point of intersection of $H(y^*, \delta)$ with T. It

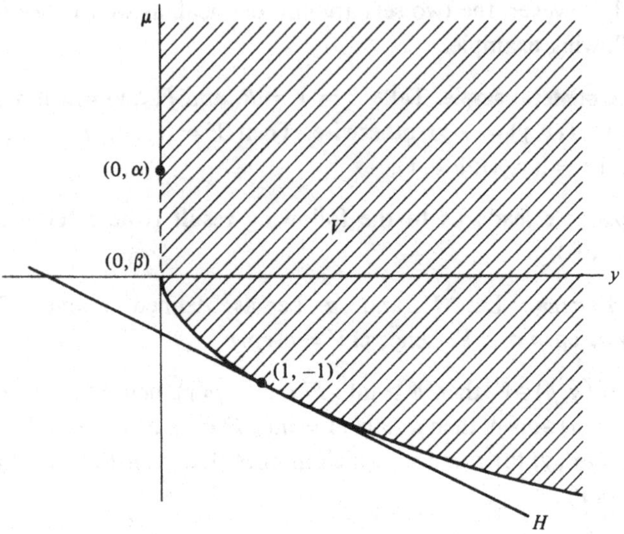

Figure 1. Geometric meaning of inf and sup.

follows, therefore, that $(0, \beta)$ is the '*highest*' point of intersection of T with hyperplanes $H(y^*, \delta)$ such that V is entirely above them. We may restrict these hyperplanes to hyperplanes *supporting* V, which is to say that $H(y^*, \delta) \cap \text{cl}\, V \neq \emptyset$, since we did not need to set $\mu + y^*y \geqslant \delta$ instead of $\mu + y^*y = \delta$.

An illustration of all this is given by the following rather pathological example, although F is convex.

3.5.9 **Example.**

Let $X = Y = R$, and
$$F(x,y) = \begin{cases} 1 & \text{if } x^2 \leqslant y \text{ and } x = 0 \\ x & \text{if } x^2 \leqslant y \text{ and } x < 0 \\ +\infty & \text{otherwise.} \end{cases}$$

Then $V = \{(y, \mu) : \mu \geqslant 1 \text{ and } y \geqslant 0, \text{ or } \mu \geqslant x \text{ and } y \geqslant x^2 \text{ for some } x < 0\}$. Figure 1 shows V, the points $(0, \alpha) = (0, 1)$ and $(0, \beta) = 0, 0)$, as well as the supporting hyperplane $H(y^* = \frac{1}{2}, \delta = -\frac{1}{2})$. The curved line is given by $y = \mu^2$ for $\mu < 0$. Notice that $(0, 1)$ belongs to V, but that $(0, 0)$ does not.

Two things are not as they should be in this example. Firstly, α is not equal to β, and secondly there is no hyperplane of the required type passing through $(0, \beta)$. The latter is tantamount to the nonexistence of a dual optimal solution y_0^*. When $\alpha > \beta$ one speaks of a (positive) *duality gap*. The example could have been worse since a primal optimal solution does exist, namely $x_0 = 0$, but this is accidental as examples exist where $\alpha > \beta$ and there is neither a primal nor a dual optimal solution (see Table 1 in 3.9).

Although in the example V has no supporting hyperplanes passing through $(0, \beta)$ which have the form $\{(y, \mu): \mu + y^*y = \delta\}$, the supporting hyperplane $\{(y, \mu): y = 0\}$ does pass through $(0, \beta)$. Its normal, however, is $(1, 0)$ and not $(y^*, 1)$ for some y^* as it should be. In the former case we speak of a *vertical hyperplane*, so that we can rephrase our interpretation of β as follows: $(0, \beta)$ is the 'highest' point of intersection of T with *non*vertical hyperplanes supporting V. Unfortunately a sequence of nonvertical hyperplanes may tend to a vertical one, and that is what happens in the example.

In the next section we shall discuss conditions under which $\alpha = \beta$, which brings us to the question of why there is a duality gap in Example 3.5.9. The reason seems to be that the set $\{(0, \mu): 0 \leqslant \mu < 1\}$ does not belong to V, although $(y, \mu) \in V$ if $0 \leqslant \mu < 1$ and $y > 0$. This leads us to compare $\mathrm{cl}(T \cap V)$ with $T \cap \mathrm{cl}\, V$. It is always true that $\mathrm{cl}(T \cap V) \subset T \cap \mathrm{cl}\, V$, but there are cases where

$$T \cap \mathrm{cl}\, V \not\subseteq \mathrm{cl}(T \cap V),$$

and these we want to regard as *abnormal*.

3.5.10 **Definition.** Given any bifunction $F: X \times Y \to [-\infty, +\infty]$ the problem of finding $\inf_X F(x, 0)$ as well as F itself is called *normal* if $\mathrm{cl}(T \cap V) = T \cap \mathrm{cl}\, V$, where T and V are as in (3.5.3) and (3.5.4).

Later on we shall come across a slightly different version of the definition of normality (see 4.3).

It would seem, at least if F is convex, that normality implies the equality of α and β. But this is not so, as the following example shows.

3.5.11 **Counterexample.** Let $X = Y = R_2$, $x = (\xi_1, \xi_2)$, $y = (\eta_1, \eta_2)$, and $F(x, y) = \xi_1 + \xi_2$ if $\xi_1 - \xi_2 + 1 \leqslant \eta_1$ and $-\xi_1 + \xi_2 + 1 \leqslant \eta_2$, and $F(x, y) = +\infty$ otherwise. It is easily shown that F is normal since V is closed. But $\alpha = +\infty$ and $\beta = -\infty$.

This counterexample is very extreme as we cannot have a larger duality gap. It will turn out that if F is convex and normal and if $\alpha > \beta$ then necessarily $\alpha = +\infty$ and $\beta = -\infty$. In other words, 'normally' convexity and normality imply the equality of α and β.

3.6 A duality theorem for convex bifunctional programming

3.6.1 **Theorem.** Let X be a linear space, Y be a locally convex topological vector space with dual Y^*, F be a convex bifunction, and let α and β be defined by $\alpha = \inf_X F(x, 0)$ and $\beta = \sup_{Y^*} \inf_{X,Y} (F(x, y) + y^*y)$. If F is not normal and $(\alpha, \beta) \neq (+\infty, -\infty)$ then $\alpha = \beta$. And if F is not normal then $\alpha > \beta$.

In almost all practical cases the condition that $(\alpha, \beta) \neq (+\infty, -\infty)$ will be satisfied, e.g. because usually $F(x, 0)$ will be finite for at least one x, so that $\alpha < +\infty$. If this condition is satisfied and if F is convex, then the theorem tells us that normality is a *necessary and sufficient* condition for α to be equal to β. Although the proof of the theorem is based on (strong) *separation*, it seems that verifying normality or deriving sufficient conditions for normality in general do not involve the relatively deep separation argument. Hence the impression is that the theorem brings the problem down to a more elementary level. This is not to say that it is easy to verify normality!

The proof relies on Lemmas 3.5.7 and 3.5.8, as well as on the next two lemmas, the first of which relies on *strong separation*.

3.6.2 **Lemma.** If $\alpha > \gamma$ for some finite value of γ and if F is convex and normal, then there exist ϵ, ρ and $y_0^* \in Y^*$, such that $\epsilon > 0$, $\rho \geq 0$ and

$$\rho\mu + y_0^* y \geq \rho\gamma + \epsilon \quad \text{if} \quad (y, \mu) \in \operatorname{cl} V. \tag{3.6.3}$$

Proof. $\alpha > \gamma$ implies that $(0, \gamma) \notin \operatorname{cl}(T \cap V)$, for otherwise every neighbourhood of $(0, \gamma)$ would contain a point $(0, \mu) \in T \cap V$, and so for some x, $\mu \geq F(x, 0) \geq \inf_X F(x, 0) = \alpha$ and we would have that $\gamma \geq \alpha$. The fact that $(0, \gamma) \notin \operatorname{cl}(T \cap V)$ and normality imply that

$(0,\gamma)\notin T\cap \mathrm{cl}\, V$, or that $(0,\gamma)\notin \mathrm{cl}\, V$. Since F is a convex function V is a convex set, as is cl V. By Theorem 3.3.2 we can therefore separate $(0,\gamma)$ and cl V strongly in the space $Y\times R$, so that we can find a $(y_0^*,\rho)\in Y^*\times R^* = Y^*\times R$ satisfying (3.6.3) for some $\epsilon>0$. In particular it follows from (3.6.3) that $\rho\mu+y_0^*y$ is bounded below for all $(y,\mu)\in \mathrm{cl}\, V$. In view of the fact that $(y,\mu+\lambda)\in V$ for all positive λ if $(y,\mu)\in V$ it follows from this that ρ cannot be negative.

Notice that we have not shown that ρ is positive, hence the hyperplane $\{(y,\mu)\colon \rho\mu+y_0^*y=\rho\gamma+\epsilon\}$ may be vertical. In the following lemma we assume that this is not the case.

3.6.4 **Lemma.** If in (3.6.3) $\rho=1$, hence if

(3.6.5) $$\mu+y_0^*y\geqslant\gamma+\epsilon \quad \text{for} \quad (y,\mu)\in \mathrm{cl}\, V,$$

then $\beta\geqslant\gamma+\epsilon$.

Proof. By the definition of β and Lemma 3.5.8 we have that $\beta\geqslant \inf_{X,Y}(F(x,y)+y_0^*y)=\inf\{\mu+y_0^*y\colon (y,\mu)\in V\}$, which by (3.6.5) is not less than $\gamma+\epsilon$.

Proof of Theorem 3.6.1

(A) First let F be normal and assume that $(\alpha,\beta)\neq(+\infty,-\infty)$. Since always $\alpha\geqslant\beta$ it is sufficient to show that $\alpha\leqslant\beta$.

Case 1. Let both α and β be finite and suppose that $\alpha>\beta$. In Lemma 3.6.2 take $\gamma=\beta$; then it follows that for some $\epsilon>0$, $\rho\geqslant 0$, y_0^*,

(3.6.6) $$\rho\mu+y_0^*y\geqslant\rho\beta+\epsilon \quad \text{if} \quad (y,\mu)\in \mathrm{cl}\, V.$$

But $\alpha=\inf_X F(x,0)$, hence for all $\epsilon'>0$ there exists an x' such that $\alpha+\epsilon'\geqslant F(x',0)$ so that $(0,\alpha+\epsilon')\in V$ and $(0,\alpha)\in \mathrm{cl}\, V$. Substitution of $(0,\alpha)$ for (y,μ) in (3.6.6) yields $\rho\alpha\geqslant\rho\beta+\epsilon$, and so necessarily $\rho>0$. As (y_0^*,ρ) is defined up to a positive factor we may take $\rho=1$, so that by Lemma 3.6.4 $\beta\geqslant\beta+\epsilon$, which is impossible. Hence $\alpha>\beta$ cannot be true.

Case 2. Let α be finite and $\beta=-\infty$. For γ take any finite value smaller than α and reason as in case 1. Then the conclusion will be that $\beta\geqslant\gamma+\epsilon$, which again is impossible.

Case 3. Let $\alpha=+\infty$ and suppose that β is finite. Take $\gamma=\beta$, as in case 1, and reason as there. Then again (3.6.6) holds for some $\epsilon > 0$, $\rho \geqslant 0, y_\circ^*$. Instead of showing that $\rho > 0$ we now show that $\rho = 0$. Suppose that $\rho > 0$, then we may take $\rho = 1$, and Lemma 3.6.4 leads to $\beta \geqslant \beta + \epsilon$, which is false. Hence indeed $\rho = 0$, so that

$$(3.6.7) \qquad y_\circ^* y \geqslant \epsilon \quad \text{if} \quad (y, \mu) \in \operatorname{cl} V.$$

This involves a *vertical* hyperplane, which we want to 'tilt' in order to be able to apply Lemma 3.6.4 again to arrive at a contradiction. To this end we consider the definition of β again and conclude that a $y^* \in Y^*$ exists such that $\inf_{X, Y}(F(x, y) + y^*y) \geqslant \beta - 1$, or by Lemma 3.5.8 that

$$(3.6.8) \qquad \mu + y^*y \geqslant \beta - 1 \quad \text{if} \quad (y, \mu) \in V,$$

which involves a *nonvertical* hyperplane. Now take $\gamma > \beta - 1$ and define ω by $\omega = (\gamma - \beta + 1)/\epsilon$; then from (3.6.7) and (3.6.8) it follows that $\omega y_\circ^* y + \mu + y^*y \geqslant \omega\epsilon + \beta - 1$ if $(y, \mu) \in V$, hence that

$$(3.6.9) \qquad \mu + (y^* + \omega y_\circ^*) y \geqslant \gamma \quad \text{if} \quad (y, \mu) \in V,$$

which involves the 'tilted' hyperplane we had in mind. Lemma 3.6.4 with $\gamma + \epsilon$ replaced by γ and with $y_\circ^*$ replaced by $y^* + \omega y_\circ^*$ implies that $\beta \geqslant \gamma$. Taking γ arbitrarily large we find from this that $\beta = +\infty$, contradicting the assumed finiteness of β. Since in this case $\beta = -\infty$ is excluded we must have that $\beta = +\infty$.

(B) Let F not be normal. Then there exists a finite γ such that $(0, \gamma) \notin \operatorname{cl}(T \cap V)$ and $(0, \gamma) \in T \cap \operatorname{cl} V$. The latter implies that $V \neq \emptyset$. We have already seen in 3.5 that Lemma 3.5.7 implies that $\alpha = \inf\{\mu : (0, \mu) \in \operatorname{cl}(T \cap V)\}$, hence $\alpha > \gamma$. When $\beta = -\infty$ it trivially follows from this that $\alpha > \beta$. So let us assume that $\beta > -\infty$. Consider the hyperplane $H(y^*, \delta) = \{(y, \mu) : \mu + y^*y = \delta\}$. In particular take $\delta = \inf_{X, Y}(F(x, y) + y^*y)$; then $\delta < +\infty$ because $V \neq \emptyset$. Moreover $\beta = \sup_{Y^*} \delta$, and since $\beta > -\infty$ we can choose y^* so that δ is finite. Then $H(y^*, \delta)$ separates $(0, \delta)$ from V, hence from $T \cap \operatorname{cl} V$, hence from $(0, \gamma)$, so that $\gamma \geqslant \delta$, or $y \geqslant \sup_{Y^*} \delta = \beta$ and again we find that $\alpha > \beta$.

3.7 Some sufficient conditions for normality

An obvious sufficient condition for normality is given in the following theorem.

3.7.1 **Theorem.** If V defined by (3.5.4) is closed then F is normal.

The proof is an immediate consequence of the definition of normality. As an example consider (finite-dimensional) linear programming:

$$F(x,y) = cx \quad \text{if} \quad Ax+y \geqslant b \quad \text{and} \quad x \geqslant 0,$$

and $F(x,y) = +\infty$ otherwise, $x \in R_n$, $y \in R_m$. Then

$$V = \{(y,\mu)\colon \mu \geqslant cx, Ax+y \geqslant b \quad \text{for some} \quad x \geqslant 0\}$$

and it can be shown that this is a closed set (see Appendix D).

The condition that V is closed does not exclude the possibility of vertical hyperplanes supporting V. For simply take $F(x,y) = x$ if $x^2 \leqslant y$ and $F(x,y) = +\infty$ otherwise, $x, y \in R$; then

$$V = \{(y,\mu)\colon \mu \geqslant x,\ y \geqslant x^2 \text{ for some } x\},$$

so that V is closed and even convex, but no nonvertical supporting hyperplane is 'optimal'. The next theorem tells us more about vertical hyperplanes.

3.7.2 **Theorem.** If $T \cap \operatorname{ri} V \neq \varnothing$ and V is convex then F is normal.

Proof. The proof is in three parts.

(A) First we show that $V \subset b+L(V)$ if $b \in \operatorname{cl} V$ (see Definition 3.3.4 for details about ri and $L(V)$). Suppose, to the contrary, that $b \in \operatorname{cl} V$ and that $v \notin b+L(V)$ for some $v \in V$. Since $L(V)$ is closed the latter implies that $(v+U) \cap (b+L(V)) = \varnothing$ for some neighbourhood U of the origin. Since $-U$ is also a neighbourhood (requirement 2 of Definition 3.2.1 for locally convex topological vector spaces) and $b \in \operatorname{cl} V$, we have that $(b-U) \cap V \neq \varnothing$, so that $b-u = v'$ for some $u \in U$ and some $v' \in V$, or $v+u = b+(v-v')$. But $v-v' \in L(V)$, hence $(v+U) \cap (b+L(V))$ is empty which is a contradiction.

(B) Now let $a \in \operatorname{ri} V$ and $b \in \operatorname{cl} V$. Then $\lambda a+(1-\lambda)b \in V$ if $0 < \lambda < 1$. This can be seen as follows. By definition of ri,

$a+U \cap L(V) \subset V$ for some neighbourhood U of the origin. Let $W = (\gamma U) \cap L(V)$ with $\gamma = -\lambda/(1-\lambda)$. Since $b \in \text{cl}\, V$, $v \in b+\gamma U$ for some $v \in V$ and by part (A) $v \in b+L(V)$, so that $v-b = w$ for some $w \in W$, which means that $b+w \in V$. Further let $u = w/\gamma$; then $u \in U \cap L(V)$ because $L(V)$ is a linear subspace, hence $a+u \in V$. Finally, by the convexity of V we have that

$$\lambda a+(1-\lambda)b = \lambda(a+u)+(1-\lambda)(b+w) \in V.$$

(C) Let us now consider the main assumption, i.e. $T \cap \text{ri}\, V \neq \emptyset$, which implies that $(0,\mu') \in T \cap \text{ri}\, V$ for some μ'. Take any $(0,\mu'') \in T \cap \text{cl}\, V$. Then by part (B) the set S defined by

$$S = \{(y,\mu): (y,\mu) = \lambda(0,\mu')+(1-\lambda)(0,\mu'') \quad \text{for some} \quad \lambda, 0 < \lambda < 1\}$$

is a subset of V. Trivially, $S = T \cap S$ hence $S \subset T \cap V$ and taking λ sufficiently small we see that $(0,\mu'') \in \text{cl}\, S \subset \text{cl}\,(T \cap V)$. As $(0,\mu'')$ was taken arbitrarily in $T \cap \text{cl}\, V$ it follows that $T \cap \text{cl}\, V \subset \text{cl}\,(T \cap V)$ and since $\text{cl}\,(T \cap V) \subset T \cap \text{cl}\, V$ always holds we must have that $\text{cl}\,(T \cap V) = T \cap \text{cl}\, V$.

If $\text{ri}\, V = \text{int}\, V$ then it is easily shown that the condition $T \cap \text{int}\, V \neq \emptyset$ excludes the possibility of vertical supporting hyperplanes $\{(y,\mu): y^*y = 0\}$, $y^* \neq 0$. This conclusion may not be drawn if only $T \cap \text{ri}\, V \neq \emptyset$, as is exemplified by the following, somewhat silly, problem where $F(x,y) = x$ if $x = 0$, $F(x,y) = +\infty$ otherwise, $x, y \in R$.

In the finite-dimensional case we have the next result whose proof is straightforward.

3.7.3 **Theorem.** Let $f, g_1, \ldots, g_k$ be convex continuous functions from a locally convex topological vector space X to R and let C be a convex compact subset of X. Then the problem of finding $\alpha = \inf\{f(x): g_i(x) \leqslant 0, \text{ for all } i, x \in C\}$ is normal if α is finite.

3.8 On the existence of optimal solutions

So far we have only considered the equality of α and β, that is of inf and sup, but now we turn to conditions that guarantee the solvability of the primal and/or the dual problem. Since F can take infinite values we must slightly reformulate the definition of optimal solution. We shall say that x_o is an (optimal) *primal solution* and that the primal problem is *solvable* if $F(x_o, 0) = \inf_X F(x, 0) < +\infty$. The primal problem is called *infeasible* if $\inf_X F(x, 0) = +\infty$. Likewise y_o^* is an (optimal) *dual solution* and the dual problem is *solvable* if

$$\inf_{X,Y}(F(x,y)+y_o^*y) = \sup_{Y^*}\inf_{X,Y}(F(x,y)+y^*y) > -\infty,$$

and the dual problem is *infeasible* if the latter expression is equal to $-\infty$.

In any practical situation $F(x, 0)$ will be nowhere $-\infty$, hence usually $\alpha = F(x_o, 0)$ will be finite. Indeed, in many theorems it is required that α is finite.

3.8.1 **Theorem.** The primal problem of finding $\alpha = \inf_X F(x, 0)$ is solvable if α is finite and if $T \cap V$ is closed, where T and V are as before:

(3.8.2) $T = \{(y, \mu): \mu \in R, y = 0 \in Y\}$

(3.8.3) $V = \{(y, \mu): \mu \in R, y \in Y, \mu \geqslant F(x, y) \quad \text{for some } x \in X\}$.

Conversely, if the primal problem is solvable then $\alpha < +\infty$ and $T \cap V$ is closed.

Proof. Always $\alpha = \inf\{\mu: (0, \mu) \in T \cap V\}$. If α is finite and $T \cap V$ is closed then $(0, \alpha) \in T \cap V$, hence $\alpha \geqslant F(x_o, 0)$ for some x_o; but of course $\alpha \leqslant F(x_o, 0)$ hence $\alpha = f(x_o, 0)$. If on the other hand $\alpha = F(x_o, 0) < +\infty$ then $T \cap V = \{(0, \mu): \mu \geqslant f(x_o, 0)\}$ so that $T \cap V$ is closed.

3.8.4 **Corollary.** If $F(x, 0) > -\infty$ for all x then the solvability of the primal problem is equivalent to α being finite and $T \cap V$ being closed.

A characterization of the existence of dual solutions requires a little more effort. We begin with a definition.

3.8.5 **Definition.** If V is nonempty, the *cone* (with apex at the origin) *generated by V at* $(0, \gamma)$ is defined by

(3.8.6*a*) $K(\gamma) = \{k: (0, \gamma) + \lambda k \in V \quad \text{for some } \lambda > 0\} \cup \{(0, 0)\}$, or equivalently by

(3.8.6b)
$$K(\gamma) = \{k: k = \lambda(v-(0,\gamma)) \quad \text{for some} \quad \lambda \geqslant 0 \quad \text{and some } v \in V\}.$$

Trivially $V \subset (0,\gamma)+K(\gamma)$. Further it is easy to show that if V is convex then so is $K(\gamma)$.

The next theorem shows that the existence of dual solutions is closely related to whether or not $(0,\beta)+\mathrm{cl}\,K(\beta)$ includes T, which is the μ-axis of the (y,μ)-space. Intuitively this is clear, for if $(0,\beta)+\mathrm{cl}\,K(\beta) \supset T$ it would seem that the only supporting hyperplane through $(0,\beta)$ is a vertical one, whereas in the opposite case there are at least some nonvertical ones. In the latter case we are best off if *all* supporting hyperplanes through $(0,\beta)$ are nonvertical. This would require relatively strong conditions, however, such as that $T \cap \mathrm{int}\,V \neq \varnothing$, which we considered in Theorem 3.7.2. But we are not concerned with that here.

3.8.7 **Theorem.** If $\beta = \sup_{Y^*} \inf_{X,Y} (F(x,y)+y^*y)$ is finite and if V is convex, then the dual problem is solvable if and only if

$$(0,\beta)+\mathrm{cl}\,K(\beta)) \not\supset T. \tag{3.8.8}$$

Proof. The proof is based on *strong separation*, as was the proof of Theorem 3.6.1. Let V be convex and let β be finite. Then V is not empty.

(A) Let $(0,\beta)+\mathrm{cl}\,K(\beta) \not\supset T$. Then $(0,\gamma) \notin (0,\beta)+\mathrm{cl}\,K(\beta)$ for some γ and Theorem 3.3.2 implies the existence of $\epsilon > 0$, ρ and $y_0^* \in Y^*$ such that

$$\rho\mu + y_0^* y \geqslant \rho\gamma + \epsilon \quad \text{if} \quad (y,\mu) \in (0,\beta)+\mathrm{cl}\,K(\beta). \tag{3.8.9}$$

Since $V \subset (0,\beta)+\mathrm{cl}\,K(\beta)$ and $(y,\mu+\lambda) \in V$ if $(y,\mu) \in V$ and $\lambda > 0$ we have (as we have seen before) that $\rho \geqslant 0$. Taking $(y,\mu) = (0,\beta)$ we find from (3.8.9) that $\rho\beta \geqslant \rho\gamma + \epsilon$, hence we further have that $\rho > 0$. Setting $\rho = 1$ we obtain

$$\mu + y_0^* y \geqslant \gamma + \epsilon \quad \text{if} \quad (y,\mu) \in (0,\beta)+\mathrm{cl}\,K(\beta). \tag{3.8.10}$$

From this it follows that $\beta \leqslant \inf\{\mu + y_0^* y: (y,\mu) \in V\}$, for otherwise $\mu' + y_0^* y' < \beta - \delta$ for some $\delta > 0$ and some $(y',\mu') \in V$. But by the second definition of $K(\beta)$,

$$\lambda(y',\mu')+(1-\lambda)(0,\beta) \in (0,\beta)+\mathrm{cl}\,K(\beta) \quad \text{if} \quad \lambda > 0,$$

so that (3.8.10) leads to $\lambda\mu' + (1-\lambda)\beta + y_0^*(\lambda y') \geqslant \gamma + \epsilon$ and this, combined with $\mu' + y_0^* y' < \beta - \delta$, gives $\beta - \lambda\delta \geqslant \gamma + \epsilon$, which cannot be true if we take λ sufficiently large. Hence we have indeed that $\beta \leqslant \inf\{\mu + y_0^* y : (y, \mu) \in V\}$, which by Lemma 3.5.8 is equal to $\inf_{X,Y}(F(x,y) + y_0^* y)$, so that $\beta \leqslant \inf_{X,Y}(F(x,y) + y_0^* y)$ and we may, of course, put the equality sign here, which shows that y_0^* is a dual solution.

(B) If, conversely, $\beta = \inf\{\mu + y_0^* y : (y, \mu) \in V\}$ for some y_0^*, then the hyperplane $\{(y,\mu) : \mu + y_0^* y = \beta\}$ separates $(0, \beta)$ and V weakly. And this implies that $(0,0)$ and $\mathrm{cl}\, K(\beta)$ can be separated. For take any $(y,\mu) \in K(\beta)$, $(y, \mu) \neq (0,0)$. Then by the first definition of $K(\beta)$, $(0,\beta) + \lambda(y,\mu) \in V$ for some $\lambda > 0$, so that $(\beta + \lambda\mu) + y_0^*(\lambda y) \geqslant \beta$; hence $\mu + y_0^* y \geqslant 0$. Since y_0^* is continuous we may also take $(y,\mu) \in \mathrm{cl}\, K(\beta)$ and still have that $\mu + y_0^* y \geqslant 0$. Therefore, e.g. $(0,-1) \notin \mathrm{cl}\, K(\beta)$, which means that $(0,\beta) + \mathrm{cl}\, K(\beta) \not\supset T$.

Notice how V was enlarged to the cone $(0,\beta) + K(\beta)$ and how the cone property of $K(\beta)$ was used.

Considering $K(\alpha)$ instead of $K(\beta)$ leads to the next theorem.

3.8.11 **Theorem.** If α is finite and if V is convex, then $\alpha = \beta$ *and* the dual problem problem is solvable if and only if

$$(0,\alpha) + \mathrm{cl}\, K(\alpha) \not\supset T. \tag{3.8.12}$$

Proof. Let α be finite and let V be convex.

(A) If $(0,\alpha) + \mathrm{cl}\, K(\alpha) \not\supset T$ we follow the reasoning of part (A) of the proof of the preceding theorem and find that

$$\alpha \leqslant \inf\{\mu + y_0^* y : (y,\mu) \in V\} \quad \text{for some} \quad y_0^*$$

so that $\alpha \leqslant \beta$ and, since $\alpha \geqslant \beta$ always, it follows that $\alpha = \beta$.

(B) If $\alpha = \beta = \inf\{\mu + y_0^* y : (y,\mu) \in V\}$ for some y_0^*, then $(0,\alpha) + \mathrm{cl}\, K(\alpha) \not\supset T$ is an immediate consequence of part (B) of the said proof.

Clearly (3.8.12) is an important condition, because the only thing it does not give is the solvability of the primal problem (assuming α finite and V convex). If it holds, one sometimes says the optimization problem involved is *stably set*. It also yields a third sufficient condition for normality. For if α is finite and if (3.8.12) holds, then the problem is normal, no matter whether V is convex or not. The proof of this statement, which is best given indirectly, is left to the reader. This result does not seem to be terribly interesting, since normality is usually combined with convexity, and then it is a trivial consequence of Theorems 3.8.11 and 3.6.1.

If α is finite then, say, $(0, \alpha-1)$ is not an element of $(0,\alpha)+K(\alpha)$; for otherwise $(0,-1) \in K(\alpha)$, hence $(0,\alpha-\lambda) \in V$ for some $\lambda > 0$, and $F(x,0) \leqslant \alpha-\lambda$ for some x, which contradicts the definition of α. Thus we have:

3.8.13 **Theorem.** If α is finite and if $K(\alpha)$ is closed, then $(0,\alpha)+\text{cl}\, K(\alpha) \not\supset T$.

Unfortunately, for $K(\alpha)$ to be closed, it is not sufficient that V is, as we see if we again take $F(x,y) = x$ if $x^2 \leqslant y$ and $F(x,y) = +\infty$ otherwise, $x, y \in R$. On the other hand the conditions of Theorem 3.7.2, which are sufficient for normality, are almost sufficient for (3.8.12) to hold.

3.8.14 **Theorem.** If $T \cap \text{int}\, V \neq \varnothing$, if α is finite and if V is convex, then $(0,\alpha)+\text{cl}\, K(\alpha) \not\supset T$.

Proof. The proof is based on *weak separation*. $T \cap \text{int}\, V \neq \varnothing$ implies that $(0,\mu') \in \text{int}\, V$ for some μ'. Further $(0,\alpha)$ is a boundary point of V, because $(0,\alpha-\epsilon) \notin V$ for all $\epsilon > 0$, which follows directly from the definitions of α and V. Let $W = \{(0,\alpha)\}$; then $\text{int}\, V \neq \varnothing$, $W \neq \varnothing$ and $\text{int}\, V \cap W = \varnothing$. Hence by Theorem 3.3.3 there exist ρ and y^* such that $(y^*,\rho) \neq (0,0)$ and such that $\rho\mu + y^*y \geqslant \rho\alpha$ if $(y,\mu) \in V$. As in previous proofs it follows that $\rho \geqslant 0$ and we can further show that $\rho > 0$. For $\rho = 0$ would lead to $y^*y \geqslant 0$ for all $(y,\mu) \in V$, which means that $y^*y \geqslant 0$ for all y in a neighbourhood of the origin of Y, since $(0,\mu') \in \text{int}\, V$. This implies (by requirement 5 of Definition 3.2.1 of a locally convex topological vector space) that $y^* = 0$, which contradicts $(y^*,\rho) \neq (0,0)$. Hence indeed $\rho > 0$ and taking $\rho = 1$ it follows that $\mu + y^*y \geqslant \alpha$ if $(y,\mu) \in V$. But since if $(y,\mu) \in (0,\alpha)+K(\alpha)$ either $(y,\mu) = (0,\alpha)$ or $\lambda(y,\mu)+(1-\lambda)(0,\alpha) \in V$ for some $\lambda > 0$, it even follows that $\mu + y^*y \geqslant \alpha$ if $(y,\mu) \in (0,\alpha)+K(\alpha)$, and hence also if$(y,\mu)\in(0,\alpha)+\mathrm{cl}\,K(\alpha)$. As $(y,\mu)=(0,\alpha-1)$ does not satisfy $\mu+y^*y\geqslant\alpha$ we see that $(0,\alpha-1)\notin(0,\alpha)+\mathrm{cl}\,K(\alpha)$, so that $(0,\alpha)+\mathrm{cl}\,K(\alpha)\not\supset T$.

If Y is a Banach space we can replace int by ri.

3.8.15 **Theorem.** If $T\cap\mathrm{ri}\,V\neq\varnothing$, if α is finite, if V is convex, and if Y is a Banach space, then $(0,\alpha)+\mathrm{cl}\,K(\alpha)\not\supset T$.

Proof. The proof is almost identical to that of the preceding theorem. The main difference is that we cannot use Theorem 3.3.3, but have to use Theorem 3.3.6 instead. Clearly $\mathrm{ri}\,V\neq\varnothing$, and $(0,\alpha)\notin\mathrm{ri}\,V$, because $T\subset L(V)$. Taking again $W=\{(0,\alpha)\}$ we have $\mathrm{ri}\,W=W\neq\varnothing$, and $\mathrm{ri}\,V\cap\mathrm{ri}\,W=\varnothing$. Since $L(V)$ is closed, so is $L(V)+L(W)$, because $L(W)=\{0\}$. Hence Theorem 3.3.6 indeed applies. We can even separate V and W in the subspace $L(V)$, as follows from part (B) of the proof of Theorem 3.3.6 (see Appendix B), because $v=(0,\alpha+1)\in V$ and $w=(0,\alpha)\in W$, so that $w-v=(0,-1)\in L(V)$. From this it follows that we can again show that $\rho>0$, because ri V in $Y\times R$ is the same as int V in $L(V)$.

Another sufficient condition for (3.8.12) to hold is related to the boundedness of the perturbation function p in the neighbourhood of the origin

3.8.16 **Theorem.** If α is finite and the perturbation function $p(y)=\inf_X F(x,y)$ is convex and bounded above in a neighbourhood of the origin, then $(0,\alpha)+\mathrm{cl}\,K(\alpha)\not\supset T$.

Proof. If α is finite and if $(0,\alpha)+\mathrm{cl}\,K(\alpha)\supset T$, then $(0,-1)\in\mathrm{cl}\,K(\alpha)$, hence each neighbourhood of $(0,-1)$ contains an element $\lambda((y,\mu)-(0,\alpha))$ for some $\lambda>0$ and some $(y,\mu)\in V$, as follows from the definition of $K(\alpha)$. If we take this neighbourhood sufficiently small, it follows that $\lambda(\mu-\alpha)$ must be negative, and hence that $\mu=\alpha-\delta$ for some $\delta>0$. From the convexity of p it follows that

$$p(-y)\geqslant(p(0)-(1-\lambda')p(y))/\lambda' \quad \text{if} \quad 0<\lambda'<1.$$

Further, $p(0)=\alpha$ and $(y,\mu)\in V$ means that $\mu\geqslant p(y)$, so that $p(y)\leqslant\alpha-\delta$. Hence

$$p(-y)\geqslant(\alpha-(1-\lambda')(-\delta))/\lambda'=\alpha+\delta(1-\lambda')/\lambda',$$

which can be made arbitrarily large by taking λ' sufficiently close to zero. Therefore p is not bounded above in any small neighbourhood of the origin. If we reverse the argument the theorem immediately follows.

The condition that p is bounded in a neighbourhood of the origin is stronger than the condition that $T \cap \text{int } V \neq \varnothing$, for if $p(y) < M$ for some M and y in some neighbourhood U of the origin then $(y, \mu) \in V$ if $y \in U$ and $\mu \geqslant M$. This is in agreement with the fact that in the proof of Theorem 3.8.14 we had to rely on a separation argument, whereas the proof of Theorem 3.8.16 is elementary.

3.9 Summary of results; examples

Let us summarize some results of the last two sections for the case when V is *convex*.

(A) First assume that α is finite.

(1) The problem is normal if and only if $\alpha = \beta$.

(2) $(0, \alpha) + \text{cl } K(\alpha) \not\supset T$ if and only if $\alpha = \beta$ and the dual problem is solvable.

(3) The problem is normal and $T \cap V$ is closed if and only if $\alpha = \beta$ and the primal problem is solvable.

(4) $(0, \alpha) + \text{cl } K(\alpha) \not\supset T$ and $T \cap V$ is closed if and only if $\alpha = \beta$ and both the primal and the dual problem are solvable.

(B) If $\alpha = -\infty$ then, of course, $\beta = -\infty$ and hence the problem is normal, but the condition $(0, \alpha) + \text{cl } K(\alpha) \not\supset T$ has no meaning. By definition the dual problem is infeasible, and if $F(x, 0) > -\infty$ for all x then the primal problem is not solvable.

(C) If $\alpha = +\infty$ then by definition the primal problem is infeasible. If V is empty then $\beta = +\infty$ and any y_0^* is a dual solution. If V is not empty we have three subcases: if $\beta = +\infty$ then the dual problem is not solvable, if β is finite then the dual problem may or may not be solvable, and if $\beta = -\infty$ then again by definition the dual problem is infeasible.

Altogether, if $F(x, 0) > -\infty$ for all x, we can distinguish sixteen cases, which are listed in Table 1, complete with an example for each.

Table 1. *Sample problems for the various cases*

Case	α	β	$\alpha = \beta$	x_o	y_0^*	X	$f(x)$	$g(x)$	C
1	fi	fi	yes	yes	yes	R	ξ_1	ξ_1	$\xi_1 \geqslant 0$
2	fi	fi	yes	yes	no	R	ξ_1	ξ_1^2	$\xi_1 \leqslant 0$
3	fi	fi	yes	no	yes	R	ξ_1	0	$\xi_1 > 0$
4	fi	fi	yes	no	no	R_2	$\xi_1 + \xi_2$	ξ_1^2	$\xi_1 \leqslant 0, \xi_2 > 0$
5	fi	fi	no	yes	yes	R	1 if $\xi_1 = 0$ ξ_1 if $\xi_1 > 0$	ξ_1	$\xi_1 \geqslant 0$
6	fi	fi	no	yes	no	R	1 if $\xi_1 = 0$ ξ_1 if $\xi_1 < 0$	ξ_1^2	$\xi_1 \leqslant 0$
7	fi	fi	no	no	yes	R_2	$1 + \xi_2$ if $\xi_1 = 0$ $\xi_1 + \xi_2$ if $\xi_1 > 0$	ξ_1	$\xi_1 \geqslant 0, \xi_2 > 0$
8	fi	fi	no	no	no	R_2	$1 + \xi_2$ if $\xi_1 = 0$ $\xi_1 + \xi_2$ if $\xi_1 < 0$	ξ_1^2	$\xi_1 \leqslant 0, \xi_2 > 0$
9	fi	$-\infty$	no	yes	no	R_2	1 if $\xi_1 = 0$ $\xi_1 - \xi_2$ if $\xi_1 > 0$	ξ_1	$\xi_1 \geqslant 0, \xi_2 \geqslant 0$
10	fi	$-\infty$	no	no	no	R_3	$1 + \xi_2$ if $\xi_1 = 0$ $\xi_1 + \xi_2 - \xi_3$ if $\xi_1 > 0$	ξ_1	$\xi_1 \geqslant 0, \xi_2 > 0, \xi_3 \geqslant 0$
11	$-\infty$	$-\infty$	yes	no	no	R	ξ_1	ξ_1	$\xi_1 \leqslant 0$
12	$+\infty$	$+\infty$	yes	no	yes	R			$C = \varnothing$
13	$+\infty$	$+\infty$	yes	no	no	R	ξ_1	ξ_1	$\xi_1 \geqslant 1$
14	$+\infty$	fi	no	no	yes	R	ξ_1	ξ_1	$\xi_1 > 0$
15	$+\infty$	fi	no	no	no	R	ξ_1	ξ_1^2	$\xi_1 < 0$
16	$+\infty$	$-\infty$	no	no	no	R_2	$\xi_1 - \xi_2$	ξ_1	$\xi_1 > 0, \xi_2 \geqslant 0$

fi, finite; x_o means primal solvable, y_0^* means dual solvable

All examples are of the type $F(x, y) = f(x)$ if $g(x) \leqslant y$ and $x \in C$, and $F(x, y) + = \infty$ otherwise. In all cases $Y = R$ (with the ordinary positive cone) and $X = R_i$ with i at most 3.

3.10 Saddle-points, saddle-values and the meaning of the Langrangian

We have already seen in 2.5 that if $\alpha = \beta$ and if both the primal and the dual problem are solvable, that is if for some x_o and some y_0^*,

$$\alpha = F(x_o, 0) = \beta = \inf_{X,Y}(F(x, y) + y_0^* y),$$

then

$$L(x_o, y^*) \leqslant L(x_o^*, y^*) \leqslant L(x, y_o^*) \tag{3.10.1}$$

if $x \in X$ and $y^* \in Y^*$, where, by definition,

$$L(x, y^*) = \inf_Y (F(x, y) + y^* y) \tag{3.10.2}$$

is the Lagrangian function, or simply the Langrangian.

If merely $\alpha = \beta$, that is if

$$\inf_X F(x, 0) = \sup_{Y^*} \inf_{X,Y} (F(x, y) + y^* y),$$

then by a similar argument to the one in 2.6 we find that

$$\begin{aligned} \inf_X \sup_{Y^*} L(x, y^*) &\geqslant \inf_X F(x, 0) \\ &= \sup_{Y^*} \inf_{X,Y} (F(x, y) + y^* y) = \sup_{Y^*} \inf_X L(x, y^*), \end{aligned}$$

so that

$$\sup_{Y^*} \inf_X L(x, y^*) = \inf_X \sup_{Y^*} L(x, y^*) = \alpha = \beta. \tag{3.10.3}$$

If (3.10.1) holds we say that (x_o, y_0^*) is a *saddle-point* of the Langrangian, whereas if (3.10.3) holds we speak of $\alpha = \beta$ as the *saddle-value* of the Langrangian.

It so happens that under certain conditions the Lagrangian shares with the bifunction the property that everything is contained in it, for the simple reason that under those conditions the bifunction F can be found if only the Lagrangian is known. To see this we consider the case where $x \in X = R_n$ and $y \in Y = R_m$ and

$$\begin{aligned} F(x, y) &= f(x) \quad \text{if } g(x) \leqslant y \text{ and } x \in C \subset X \\ &= +\infty \quad \text{otherwise.} \end{aligned} \tag{3.10.4}$$

Then with $y^* \in Y^*$, the dual of Y which is R_m again,

(3.10.5) $$L(x, y^*) = f(x) + y^* g(x) \quad \text{if} \quad x \in C \quad \text{and} \quad y^* \geqslant 0$$
$$= -\infty \quad \text{if} \quad x \in C \quad \text{and} \quad y^* \ngeqslant 0$$
$$= +\infty \quad \text{if} \quad x \notin C.$$

Here by $g(x) \leqslant y$ we mean that the ith component of $g(x) - y$ is nonpositive for all $i = 1, \ldots, m$, so that the positive cone of $Y = R_m$ is what is usually called the positive orthant (or rather the nonnegative orthant) of R_m. Now F is related to L as follows.

(3.10.6) $$F(x, y) = \sup_{Y^*} (L(x, y^*) - y^* y).$$

Let us verify this: if $x \notin C$, then $L(x, y) = +\infty$, so that $F(x, y) = +\infty$. If $x \in C$ we need not take the supremum in (3.10.6) over the entire space Y^* but only over $y^* \geqslant 0$, because otherwise $L(x, y) = -\infty$. Hence if $x \in C$ we have to show that

$$F(x, y) = \sup \{f(x) + y^*(g(x) - y) \colon y^* \geqslant 0\}.$$

If $g(x) \nleqslant y$ then for some i the ith component of $g(x) - y$ is positive and taking y^* equal to zero except for its ith component, which is taken arbitrarily large, we see that $\sup = +\infty$, in agreement with $F(x, y) = +\infty$. Finally if $x \in C$ and $g(x) \leqslant y$ then the supremum is attained at $y^* = 0$ and we get $F(x, y) = f(x)$.

A similar result can be obtained if $Y = R_m$ is replaced by some locally convex topological vector space with positive cone P, as long as P is *closed*, as is the nonpositive orthant of R_m. For if P is closed we can apply Theorem 3.4.2. If again $g(x) \nleqslant y$, or $z = y - g(x) \ngeqslant 0$, then there exists a $y^* = z^* \geqslant 0$ such that $y^*(y - g(x)) < 0$, so that $\sup \{f(x) + y^*(g(x) - y) \colon y^* \geqslant 0\} = +\infty$. The rest of the reasoning is as before.

The condition that P be closed is essential, as we see from the next example.

3.10.7 **Counterexample.** Let $X = Y = R$ and $F(x, y) = -x$ if $g(x) < y$ and $x \leqslant 0$, $F(x, y) = +\infty$ otherwise, with $g(0) = 0$, $g(x) = -1$ if $x < 0$ and $g(x) = +\infty$ if $x > 0$. Then $F(x, y) = 0$ if $x = 0$ and $y > 0$, $F(x, y) = -x$ if $x < 0$ and $y > -1$, and $F(x, y) = +\infty$ otherwise. Also $L(x, y^*) = 0$ if $x = 0$ and

$y^* \geqslant 0$, $L(x,y^*) = -x - y^*$ if $x < 0$ and $y^* \geqslant 0$, $L(x,y^*) = -\infty$ if $x \leqslant 0$ and $y^* < 0$, and $L(x,y^*) = +\infty$ if $x > 0$. Denoting the right hand side of (3.10.6) by $F'(x,y)$, we obtain $F'(x,y) = 0$ if $x = 0$ and $y \geqslant 0$, $F'(x,y) = -x$ if $x < 0$ and $y \geqslant -1$, and $F'(x,y) = +\infty$ otherwise. Hence F and F' are different if $x \leqslant 0$ and $y = 0$ or $y = -1$, and this results in different perturbation functions, i.e. $p(y) = \inf\{-x\colon g(x) < y, x \leqslant 0\}$ and if we let p' be the perturbation function induced by F' then $p'(y) = \inf\{-x\colon g(x) \leqslant y, x \leqslant 0\}$, so that $p(-1) \neq p'(-1)$.

The dual bifunction, on the other hand, is simply recovered from L, since

$$F^d(y^*,x^*) = \inf_{X,Y}(F(x,y) - x^*x + y^*y) = \inf_X (L(x,y^*) - x^*x).$$

Clearly not everything is beautifully symmetric, and it seems this has something to do with sets or functions being closed. In the present case this is the fact that the positive cone is closed, but not all problems do involve a closed positive cone. For example, consider again Example 2.5.11, where the set G is not closed and $x_o = 0$ is not an optimal solution, and yet $(x_o, y_0) = (0, 1)$ is a saddle point of L. Again (3.10.6) does not hold, although in this case

$$p(y) = \inf_X \sup_{Y^*} (L(x,y^*) - y^*y = p'(y).$$

We shall not go into this problem now but postpone its treatment to 4.2 and 4.3.

The existence of saddle-points and saddle-values is very important, as becomes clear from the next two theorems.

3.10.8 **Theorem.** Let a bifunction F be given, let L be defined by (3.10.2), let it be possible to recover F from L by (3.10.6), and let (x_o, y_0^*) be a saddle-point of L. Then x_o is a primal solution and y_0^* is a dual solution, whether F is convex or not.

Proof.

$$\begin{aligned}\sup_{Y^*}\inf_X L(x,y^*) &\leqslant \inf_X \sup_{Y^*} L(x,y^*) \leqslant \sup_{Y^*} L(x_o,y^*)\\ &= L(x_o,y_0^*) \leqslant \inf_X L(x,y_0^*) \leqslant \sup_{Y^*}\inf_X L(x,y^*),\end{aligned}$$

hence we have equality everywhere. In particular

$$\sup_{Y^*}\inf_X L(x,y^*) = \inf_X L(x,y_0^*)$$

showing that y_0^* is a dual solution. Furthermore we have that

$$\inf_X F(x,0) = \inf_X \sup_{Y^*} L(x,y^*) = \sup_{Y^*} L(x_o,y^*) = F(x_o,0)$$

which means that x_o is a primal solution.

3.10.9 **Theorem.** Let everything be as in the preceding theorem, except that L has a saddle-value instead of a saddle-point. Then the primal infimum is equal to the dual supremum.

Proof. The proof is similar to the preceding one.

A word of caution regarding the convexity of F is in order here. For although it is true that no conditions are necessary as far as the convexity of F as a function of x alone is concerned, the function $F(x,\cdot)$ must be convex for each x. The latter follows from (3.10.6) since the supremum of a set of linear functions is always a convex function.

The discussion of the Langrangian is taken up again in 4.5.

3.11 Lagrangian duality

In this section we consider bifunctions F of the following type

(3.11.1) $$F(x,y) = f(x) \quad \text{if} \quad g(x) \leqslant y \quad \text{and} \quad x \in C$$
$$= +\infty \quad \text{otherwise,}$$

that is we perturb the *inequality* $g(x) \leqslant 0$ by replacing its right-hand side by y. The kind of duality that results from the choice is usually called *Lagrangian duality*, as the corresponding Lagrangian function, namely $f(x) + y^*g(x)$ (if $x \in C$ and $y^* \geqslant 0$), is the function that arises in classical problems which are solved by means of Lagrangian multipliers. The reader should be warned here, in order to avoid possible confusion, that we use the word 'Lagrangian' in 'Lagrangian function' in a much wider sense than in 'Lagrangian duality'.

3.11.2 **Theorem.** Let X be a linear space, C a convex subset of X, Y a locally convex topological vector space with positive cone P and dual Y^*, f a convex function $f\colon C \to R$, and g a convex function $g\colon C \to Y$. Assume that

(3.11.3) $\alpha = \inf\{f(x): g(x) \leqslant 0 \quad \text{and } x \in C\}$

is finite. Then there exists a $y_\circ^* \in Y^*$ such that

(3.11.4) $\inf\{f(x) + y_\circ^* g(x): x \in C\} = \alpha \quad \text{and} \quad y_\circ^* \geqslant 0$

if there exists an $\hat{x} \in C$ such that $g(\hat{x}) < 0$, or, what is the same, $-g(\hat{x}) \in \operatorname{int} P$, which implies that the interior of P must be nonempty.

Proof. Define F as in (3.11.1) and let

$$V = \{(y, \mu): \mu \in R, \mu \geqslant F(x, y) \quad \text{for some } x \in X\}.$$

Then

$$V = \{(y, \mu): \mu \in R, \mu \geqslant f(x), y \geqslant g(x) \quad \text{for some } x \in C\},$$

and $(0, f(\hat{x}) + 1) \in \operatorname{int} V$. The latter can be seen as follows. Since $-g(\hat{x}) \in \operatorname{int} P$, $-g(\hat{x}) + U \subset P$ for some neighbourhood U of the origin. Define W by

$$W = \{(y, \mu): |\mu - f(\hat{x}) - 1| < 1, y \in U\};$$

then $W \subset V$. For take any $(y, \mu) \in W$; then

$$\mu > f(\hat{x}) \text{ and } -g(\hat{x}) + y \in -g(\hat{x}) + U \subset P,$$

so that indeed $(y, \mu) \in V$. Since $W \subset V$, $(0, f(\hat{x}) + 1) \in \operatorname{int} V$.

Now it can easily be shown that V is convex, hence it follows from Theorems 3.8.14 and 3.8.11 that

$$\begin{aligned} \alpha = \beta &= \sup_{Y^*} \inf_{X,Y} (F(x, y) + y^* y) \\ &= \inf_{X,Y} (F(x, y) + y_\circ^* y) \quad \text{for some} \quad y_\circ^* \in Y^*. \end{aligned}$$

Simplifying this we find that

$$\inf_{X,Y} \{f(x) + y_\circ^* y: g(x) \leqslant y, x \in C\} = \alpha.$$

We now show that $y_\circ^* \geqslant 0$. For suppose that $y_\circ^* \ngeqslant 0$; then there would be a $y \geqslant 0$ such that $y_\circ^* y < 0$, as follows from the definition of dual positive cone, hence $\alpha \leqslant f(\hat{x}) + y_\circ^*(g(\hat{x}) + \lambda y)$ for all $\lambda > 0$, so that $\alpha = -\infty$, contrary to the assumption that α is finite. Knowing now that $y_\circ^* \geqslant 0$, we see that the infimum over Y in

$$\inf_{X,Y} \{f(x) + y_\circ^* y: g(x) \leqslant y, x \in C\}$$

is attained at $y = g(x)$ and this immediately leads to (3.11.4).

The condition that $g(\hat{x}) < 0$ for some $\hat{x} \in C$ is a *regularity condition*, which shows the peculiarity that only g is involved, not f. For this reason such a condition is termed *constraint qualification*, often abbreviated to CQ, and the present one is known as *Slater's constraint qualification.*

3.11.5 **Corollary.** If in addition to the assumptions of Theorem 3.11.2 x_o is assumed to be a primal solution, then

(3.11.6) $$y_o^* g(x_o) = 0.$$

Proof. Since $y_o^* \geqslant 0$ and $g(x_o) \leqslant 0$, we have that $y_o^* g(x_o) \leqslant 0$. And since $f(x_o) = \alpha$, it follows from (3.11.4) that $y_o^* g(x_o) \geqslant 0$.

The Lagrangian function follows from its definition, i.e.

$$L(x, y^*) = \inf_Y ((F(x, y) + y^* y),$$

and (3.11.1):

(3.11.7) $$\begin{aligned} L(x, y^*) &= f(x) + y^* g(x) && \text{if } x \in C \text{ and } y^* \geqslant 0 \\ &= -\infty && \text{if } x \in C \text{ and } y^* \not\geqslant 0 \\ &= +\infty && \text{if } x \notin C. \end{aligned}$$

3.11.8 **Example** (linear programming). Let $X = R_n$, $Y = R_m$, $f(x) = cx$, $g(x) = a - Ax$, with $c \in R_n^*$, $a \in R_m$ and A an $m \times n$ matrix, and $C = \{x\colon x \geqslant 0\}$. It follows from the theorem that if $\alpha = \inf\{cx\colon Ax \geqslant a, x \geqslant 0\}$ is finite and if $A\hat{x} > a$ for some $\hat{x} \geqslant 0$, then $\alpha = \inf\{cx + y_o^*(a - Ax)\colon x \geqslant 0\}$ for some $y_o^* \geqslant 0$. But since α is finite necessarily $c - y_o^* A \geqslant 0$, and because of this the infimum is attained at $x = 0$, so that $\alpha = y_o^* a$. The dual problem can be reduced to a very simple form, for $\beta = \sup_{Y^*}\{\inf\{cx + y^*(a - Ax)\colon x \geqslant 0\}\colon y^* \geqslant 0\} = \sup_{Y^*}\{y^* a\colon y^* A \leqslant c, y^* \geqslant 0\}$, hence the dual problem is again linear programming. If we write it as an infimum problem, i.e.

$$-\beta = \inf_{Y^*}\{-y^* a\colon -y^* A \geqslant -c, y^* \geqslant 0\},$$

and dualize again, then the result is our original problem.

It so happens that the constraint qualification in this example is superfluous. We come back to this in 3.13.

In order to get more insight into the necessity and meaning of the constraint qualification in Theorem 3.11.2, first consider a case where $Y = l_1$, the space of all absolutely convergent sequences $y = (\eta_1, \eta_2, \ldots)$ with norm $|y| = \Sigma |\eta_i|$ and let the positive cone be $P = \{y: \eta_i \geqslant 0, \text{ all } i\}$. Then we cannot apply the theorem, because $\operatorname{int} P = \varnothing$. To see this take any $y \in P$ and any $\epsilon > 0$. Then $0 \leqslant \eta_i < \frac{1}{2}\epsilon$ if i is large enough and $y' = (\eta_1, \ldots, \eta_{i-1}, -\frac{1}{2}\epsilon, \eta_{i+1}, \ldots) \notin P$; but $|y - y'| = \eta_i + \frac{1}{2}\epsilon > \epsilon$. We cannot hope to create a nonempty interior of P by changing to a weaker topology, because all that will happen is that certain open sets will cease to be open. It follows, therefore, that $Y = l_1$ is not a suitable perturbation space (or 'constraint space') if Slater's CQ is involved.

Let us now take $Y = l_\infty$, the space of all absolutely bounded sequences $y = (\eta_1, \eta_2, \ldots)$ with norm $|y| = \sup |\eta_i|$, and let P be as before. Then it is easily shown that $e = (1, 1, \ldots) \in \operatorname{int} P$, hence $\operatorname{int} P \neq \varnothing$. Another difficulty might turn up, however, because l_∞^* contains awkward elements. For example let

$$X = C = R, \quad \text{and} \quad \alpha = \inf\{-x: x \leqslant 1/i, i = 1, 2, \ldots\}.$$

Then everything is convex (and linear even) $\alpha = 0$, and, if we let

$$g(x) = (x-1, x-\tfrac{1}{2}, x-\tfrac{1}{3}, \ldots), \quad g(-1) < 0.$$

Hence by the theorem a $y_o^* \in l_\infty^*$ exists such that

$$\text{(3.11.9)} \qquad \inf\{-x + y_o^* g(x): x \in R\} = 0, \quad y_o^* \geqslant 0.$$

Unfortunately if $y = (\eta_1, \eta_2, \ldots) \in l_\infty$ we cannot compute $y_o^* y$ via an expression like $\Sigma \eta_{oi}^* \eta_i$, such that $(\eta_{o1}^*, \eta_{o2}^*, \ldots) \in l_1$, because then (3.11.9) would lead to $\Sigma \eta_{oi}^* = 1$ and $\Sigma_{oi}^*/i = 0$, which is impossible because $\eta_{oi}^* \geqslant 0$. Since $g(x) = xe - (1, \frac{1}{2}, \frac{1}{3}, \ldots)$ we have that

$$y_o^* g(x) = x \cdot y_o^* e - y_o^*(1, \tfrac{1}{2}, \tfrac{1}{3}, \ldots)$$

and it follows from (3.11.9) that $y_o^* e = 1$ and $y_o^*(1, \frac{1}{2}, \frac{1}{3}, \ldots) = 0$. The latter implies that $y_o^* y = 0$ if y is any unit vector $(0, \ldots, 0, 1, 0, \ldots)$. We come to the following strange conclusion, already mentioned in Example 3.2.9: there exists a $y_o^* \in l_\infty^*$ such that $y_o^* e = 1$, whereas $y_o^* y = 0$ for all y with only finitely many nonzero elements. Such a y_o^* is termed *purely finitely additive*, and sometimes *singular*. It is related to a bounded additive set function on the family of all subsets of the natural numbers, which is zero on every finite subset. Using the term 'singular' might be confusing since, in general, bounded additive functions can be decomposed into a continuous and a singular part with respect to a certain countably additive measure. But then 'singular' has quite another meaning, for such a singular part could well be *countably* additive. Details can be found in textbooks on functional analysis.

Forming the dual pair (l_∞, l_1) and changing to the Mackey topology would, of course, be of no help, since $y_\circ^* y$ could not be computed via a suitable bilinear form. Apparently we have to accept the fact that cases exist where either no $y_\circ^*$ exists, or $y_\circ^*$ is purely finitely additive. In the latter case we are no better off than in the former, for it seems impossible to define a purely finitely additive $y_\circ^*$ completely *and* constructively; that is, it seems impossible to write a computer program for any $y_\circ^*$ involved in our last example, such that the program will compute $y_\circ^* y$ for *any* given y. The reader may try.

The example with $\alpha = \inf\{-x : x \leqslant 1/i, \text{ all } i\}$ is, of course, somewhat abnormal, because $x \leqslant 1/i$ for all i implies that $x \leqslant 0$. If this '*induced constraint*' were added all the difficulties would disappear.

In the next theorem we do not assume that for some $\hat{x}, \hat{x} \in C$ and $-g(\hat{x}) \in \operatorname{int} P$ but that $\hat{x} \in \operatorname{ri} C$ and $-g(\hat{x}) \in P$, together with some other assumptions. It is no longer required that the positive cone P have a nonempty interior. This enables us to consider *equality* constraints by simply letting $P = \{0\}$.

3.11.10 **Theorem.** Theorem 3.11.2 remains true if we replace the condition that for some $\hat{x} \in C$, $g(\hat{x}) < 0$ by the following conditions,

(3.11.11) for some $\hat{x} \in \operatorname{ri} C$, $\quad g(\hat{x}) \leqslant 0$,

(3.11.12) f is upper semi-continuous (u.s.c.) at $\hat{x}$ when X is an infinite-dimensional topological vector space, and

(3.11.13) for all neighbourhoods U of the origin of X there exists a neighbourhood S of the origin of Y such that

$$S \subset g(\hat{x} + U) + P.$$

Proof. We may assume that $\operatorname{ri} C = \operatorname{int} C$, because if this is not the case then we may replace X by $L(C)$ (see Definition 3.3.4).

Define V as in the proof of Theorem 3.11.2. Then $(0, f(\hat{x})+1) \in \operatorname{int} V$. To see this observe that there must exist a neighbourhood U of the origin of X such that $\hat{x}+U \subset C$ and such that $f(x) < f(\hat{x})+\frac{1}{2}$ if $x \in \hat{x}+U$. This is because of (3.11.11) and (3.11.12), and because if $X = R_n$ f is automatically continuous, and hence u.s.c., at $\hat{x}$, as is shown in Appendix C. Let S be as in (3.11.13). Then

$$W = \{(y, \mu): |\mu - f(\hat{x}) - 1| < \tfrac{1}{2}, y \in S\}$$

is a neighbourhood of $(0, f(\hat{x})+1)$ which lies in V. For let $(y, \mu) \in W$ then $\mu > f(\hat{x})+\frac{1}{2}$ and $y = g(x)+y' \geqslant g(x)$ for some $x \in \hat{x}+U$ and some $y' \in P$. Hence $x \in C$ and $\mu > f(\hat{x})+\frac{1}{2} > f(x)$, so that indeed $(y, \mu) \in V$.

The remainder of the proof is identical to the last part of the proof of Theorem 3.11.2.

3.12 Lagrangian duality; affine equality constraints

In this section we replace the constraint $g(x) \leqslant 0$ by $g(x) = 0$. In fact the latter is a special case of the former, if we simply let the positive cone be $P = \{0\}$. The interior of this cone is empty, however, so that we cannot apply Theorem 3.11.2. Another trick might be to replace $g(x) = 0$ by $g(x) \leqslant 0$ and $-g(x) \leqslant 0$ and to define a suitable positive cone in $Y \times Y$, but this would not work either if one wanted to apply that theorem, because there is no $\hat{x}$ such that $g(\hat{x}) < 0$ and $-g(\hat{x}) < 0$. Yet writing $g(x) = 0$ as two inequalities has an advantage in that it suggests that g as well as $-g$ should be convex. So let us assume that g is an *affine* function. Another assumption will be that both X and Y must be Banach spaces, because we want to use the open mapping theorem.

3.12.1 **Theorem.** Let X be a Banach space, C a convex subset of X, Y another Banach space with dual Y^*, f a convex function $f: C \to R$, and g a continuous affine function $g: L(C) \to Y$, mapping $L(C)$ *onto* Y, and defined by $g(x) = Bx - b, x \in X$, with $b \in Y$ and B a continuous linear function $B: L(C) \to Y$. Assume that

(3.12.2) $$\alpha = \inf\{f(x): Bx = b, x \in C\}$$

is finite. Then there exists a $y_\circ^* \in Y^*$ such that

(3.12.3) $$\inf\{f(x)+y_\circ^*(Bx-b) \colon x \in C\} = \alpha$$

if the following regularity conditions are satisfied:

(3.12.4) $$B\hat{x} = b \quad \text{for some} \quad \hat{x} \in \operatorname{ri} C,$$
(3.12.5) f is upper semi-continuous (u.s.c.) at $\hat{x}$, if X is infinite-dimensional.

Proof. Since $L(C)$ is a closed subspace of the Banach space X, $L(C)$ itself is a Banach space. Hence instead of X we may equally well take $L(C)$. In other words, we may assume that $L(C) = X$, hence that $\operatorname{ri} C = \operatorname{int} C$.

Introduce the positive cone in $Y, P = \{0\}$; then we can invoke Theorem 3.11.10 if we can show that (3.11.13) holds, that is if for all neighbourhoods U of the origin of X there exists a neighbourhood S of the origin of Y such that $S \subset B\hat{x}-b+BU = BU$. By the *open mapping theorem* (see Appendix B) this is indeed true, and in fact we may even write $S = BU$.

When Y is finite-dimensional the condition that B should be a mapping *onto* Y is not really restrictive, because we can always take a subspace of Y, if necessary, which B does map onto.

The condition that $\hat{x} \in \operatorname{ri} C$ cannot be relaxed to $\hat{x} \in C$, as we see from the following example.

3.12.6 **Counterexample.** $X = R_2$, $x = (\xi_1, \xi_2)$, $f(x) = -\xi_1$, $Y = R$, $Bx-b = \xi_2$, $C = \{x \colon \xi_1 \geqslant 0, \xi_1^2 \leqslant \xi_2\}$. Then $B\hat{x} = b$ and $\hat{x} \in C$ imply that $\hat{x} = 0$ which is not an element of $\operatorname{int} C$, although $\operatorname{int} C \neq \varnothing$. And no $y_\circ^*$ exists. For if it did it would have to be nonnegative, as is easily shown, but

$$\inf\{-\xi_1 + y_\circ^* \xi_1^2 \colon \xi_1 \geqslant 0\} = 0$$

and this is impossible for $y_\circ^* > 0$ as well as for $y_\circ^* = 0$.

The bifunction leading to (3.12.2) is

(3.12.7) $$\begin{aligned} F(x,y) &= f(x) \quad \text{if} \quad Bx = b+y \quad \text{and} \quad x \in C \\ &= +\infty \quad \text{otherwise,} \end{aligned}$$

and the Lagrangian becomes

$$(3.12.8)\qquad \begin{aligned} L(x,y^*) &= f(x)+y^*(Bx-b) \quad \text{if} \quad x\in C \\ &= +\infty \quad\quad\quad\quad\quad\quad\;\; \text{if} \quad x\notin C. \end{aligned}$$

3.13 Lagrangian duality; mixed constraints; linear programming

Comparing the discussions of the preceding two sections we see that when dealing with equality constraints we had to assume that f was u.s.c. (upper semi-continuous) at $\hat{x}$, whereas this requirement was not necessary for inequality constraints. In the case of *mixed constraints*, involving both equality and inequality constraints, a similar extra condition is necessary. Consider the problem of finding

$$\inf\{f(x)\colon g(x)\leqslant 0, Bx=b, x\in C\}.$$

Then we can expect that, among others, the following conditions must be satisfied: $g(\hat{x})<0$, $B\hat{x}=b$ for some $\hat{x}\in \operatorname{ri} C$, and f is u.s.c. at $\hat{x}$. In addition to these we shall need that g is continuous at $\hat{x}$.

The proof of the theorem below can be given in a form similar to the proofs of Theorems 3.11.2 and 3.12.1, by invoking Theorems 3.8.14. and 3.8.11. We shall, however, follow a different line of reasoning, and apply *partial dualization*, which means that the constraints are not all perturbed together but that instead this is done in two stages. This will permit us to use Theorems 3.12.1 and 3.11.2, instead of Theorems 3.8.14 and 3.8.11. We start with a simple lemma.

3.13.1 **Lemma.** If $-g(\hat{x})\in \operatorname{int} P\subset Y$, $\hat{x}\in \operatorname{int} C\subset X$ and g is continuous at $\hat{x}$, then $x\in \operatorname{int} C'$, with $C'=\{x\colon -g(x)\in P, x\in C\}$.

Proof. For some neighbourhood S of the origin of Y we have that $-g(\hat{x})+S\subset P$, and for some neighbourhood U of the origin of X, that $\hat{x}+U\subset C$. Since g is continuous at $\hat{x}$, we can take U such that $-g(x)\in -g(\hat{x})+S$ if $x\in \hat{x}+U$. From this it immediately follows that $\hat{x}+U\subset C'$.

3.13.2 **Theorem.** Let X be a Banach space, C a convex subset of X, Y a locally convex topological vector space with dual Y^* and positive cone P (with nonempty interior), Z a Banach space with dual Z^*, f a convex function $f\colon C\to R$, g a convex function

$g: C \to Y$, and (B, b) a continuous affine function, mapping $L(C)$ *onto* Y and defined by $(B, b)x = Bx - b$, $x \in X$, $b \in Y$, where B is a continuous linear function $B: L(C) \to Y$. Assume that

$$\alpha = \inf\{f(x): g(x) \leqslant 0, Bx - b = 0, x \in C\} \tag{3.13.3}$$

is finite. Then there exist $y_0^* \in Y^*$ and $z_0^* \in Z^*$ such that

$$\inf\{f(x) + y_0^* g(x) + z_0^*(Bx - b): x \in C\} = \alpha, \quad y_0^* \geqslant 0, \tag{3.13.4}$$

if

$$g(\hat{x}) < 0, B\hat{x} = b \quad \text{for some} \quad \hat{x} \in \operatorname{ri} V, \tag{3.13.5}$$

$$f \text{ is u.s.c. at } \hat{x} \text{ if } X \text{ is infinite-dimensional and } g \text{ is continuous at } \hat{x}. \tag{3.13.6}$$

Proof. Again we can reduce the problem to one with $\operatorname{ri} C = \operatorname{int} C$ (see proof of Theorem 3.12.1). Now define C' as in the lemma. Then $\hat{x} \in \operatorname{int} C'$ and we can apply Theorem 3.12.1 with C replaced by C', Y by Z, and Y^* by Z^* Hence for some $z_0^* \in Z^*$

$$\inf\{f(x) + z_0^*(Bx - b): g(x) \leqslant 0, x \in C\} = \alpha. \tag{3.13.7}$$

The desired result now follows from Theorem 3.11.2 with $f(x)$ replaced by $f(x) + z_0^*(Bx - b)$.

We come now to a problem where we cannot apply Theorem 3.8.14, because $\operatorname{int} V$ may be nonempty (thereby ruling out the possibility of considering $\operatorname{ri} V$) but $T \cap \operatorname{int} V$ may be empty. Given such a problem and the corresponding bifunction F, so that

$$V = \{(y, \mu): \mu \geqslant F(x, y) \quad \text{for some} \quad x \in X\},$$

it can happen that apart from the desired nonvertical hyperplanes supporting V at $(0, \beta)$ there exist vertical ones *as well*. As an extremely simple case, where $X = Y = R$, consider $f(x, y) = x$ if $x \leqslant y$ and $x \geqslant 0$, and $F(x, y) = +\infty$ otherwise. Then V is the nonnegative orthant of the (y, μ)-space, $\operatorname{int} V \neq \varnothing$, $T \cap \operatorname{int} V = \varnothing$ and hyperplanes of both types exist.

In the next theorem, which is concerned with a *linear* optimization problem with *finitely many constraints*, no regularity condition is necessary at all.

3.13.8 **Theorem.** Let c be a continuous linear functional from a locally convex topological vector space X to R, A be a continuous linear function from X to $Y = R_m$ and $a \in R_m$. If

$$\alpha = \inf\{cx: Ax \geqslant a\}$$

is finite, then

$$\alpha = \min\{cx: Ax \geqslant a\} = \max\{y^*a: y^*A = c, y^* \geqslant 0\}, \tag{3.13.10}$$

which is equivalent to the existence of x_o and y_o^* such that

$$Ax_o \geqslant a, y_o^*A = c, y_o^* \geqslant 0, y_o^*(Ax_o - a) = 0. \tag{3.13.11}$$

Proof. The dual problem is that of finding

$$\beta = \sup_{y^* \geqslant 0} \inf_X (cx + y^*(a - Ax)),$$

where the infimum is not $-\infty$ only if $c - y^*A = 0$, so that

$$\beta = \sup\{y^*a: y^*A = c, y^* \geqslant 0\}.$$

Trivially (3.13.11) follows from (3.13.10) and $y_o^*g(x_o) = 0$. Conversely if x_o and y_o^* exist such that (3.13.11) is true then $cx_o = y_o^*Ax_o = y_o^*a$, hence $\alpha = \beta$ since x_o and y_o^* are feasible.

Let the bifunction be $F(x, y) = cx$ if $y \geqslant a - Ax$ and $F(x, y) = +\infty$ otherwise, and let

$$V = \{(y, \mu): \mu \geqslant cx, y \geqslant a - Ax \quad \text{for some } x\}.$$

Then

$$V - (a, 0) = \left\{(y, \mu): \begin{pmatrix} y \\ \mu \end{pmatrix} = \begin{pmatrix} -A \\ c \end{pmatrix} x + \begin{pmatrix} z \\ \tau \end{pmatrix} \quad \text{for some } x \text{ and some } \begin{pmatrix} z \\ \tau \end{pmatrix} \geqslant 0\right\},$$

so that $V - (a, 0)$ is the sum of the linear subspace

$$D = \left\{(y, \mu): \begin{pmatrix} y \\ \mu \end{pmatrix} = \begin{pmatrix} -A \\ c \end{pmatrix} x \quad \text{for some } x\right\} \quad \text{of} \quad R_{m+1}$$

and the nonnegative orthant Q of R_{m+1}. Hence, by the corollary of Appendix D, V is closed. By Theorem 3.8.1 the infimum is a minimum, and by Theorem 3.7.1 the problem is normal and

$$\alpha = \sup\{y^*a: y^*A = c, y^* \geqslant 0\}.$$

Let the infimum be attained at x_0. It remains to be shown that (3.13.11) holds for some y_0^*. Now if the ith constraint of $Ax \geqslant a$ is not active at x_0, then put the ith component of y_0^* equal to zero. This reduces the problem to one where $Ax_0 = a$, so that $y_0^*(Ax_0 - a) = 0$ is automatically satisfied. In other words it only remains to be shown that $\sup\{y^*a: y^*A = c, y^* \geqslant 0\}$ is a maximum. Consider the (primal) problem of finding $\inf\{-y^*a: y^*A = c, y^* \geqslant 0\}$. Then the corresponding 'V set' becomes

$$V^* = \{(z^*, \mu): \mu \geqslant -y^*a, z^* = y^*A - c \quad \text{for some } y^* \geqslant 0\},$$

which we can write as

$$(-c, 0) + \left\{(z^*, \mu): (z^*, \mu) = (y^*, \tau)\begin{pmatrix} A & -a \\ 0 & 1 \end{pmatrix} \quad \text{for some } (y^*, \tau) \geqslant 0\right\}.$$

By Appendix D this set is closed, hence V^* is closed and once again by Theorem 3.8.1 the infimum is a minimum and $\sup\{y^*a: y^*A = c, y^* \geqslant 0\}$ is a maximum.

When X, too, is finite-dimensional the problem of finding α is usually called *linear programming*, and the theorem is equivalent to the next two corollaries.

3.13.12 **Corollary.** Let everything be as in the theorem and let $X = R_n$. If

(3.13.13) $$\alpha = \inf\{cx: Ax \geqslant a, x \geqslant 0\}$$

is finite then

(3.13.14)
$$\alpha = \min\{cx: Ax \geqslant a, x \geqslant 0\} = \max\{y^*a: y^*A \leqslant c, y^* \geqslant 0\}.$$

Proof. In the theorem replace A by $\begin{pmatrix} A \\ I \end{pmatrix}$ where I is an identity matrix. Then $\alpha = \max\{y^*a: y^*A + z^* = c, y^* \geqslant 0, z^* \geqslant 0\}$ from which the result immediately follows.

3.13.15 **Corollary.** Let everything be as in the theorem and let $X = R_n$. If

(3.13.16) $$\alpha = \inf\{cx: Ax = a, x \geqslant 0\}$$

is finite then

(3.13.17) $\alpha = \min\{cx: Ax = a, x \geqslant 0\} = \max\{y^*a: y^*A \leqslant c\}$.

Proof. In the preceding corollary replace A by $\begin{pmatrix} A \\ -A \end{pmatrix}$ and a by $\begin{pmatrix} a \\ -a \end{pmatrix}$. Then

$$\begin{aligned}\alpha &= \max\{(y_1^* - y_2^*)\, a: (y_1^* - y_2^*) A \leqslant c, y_1^* \geqslant 0, y_2^* \geqslant 0\} \\ &= \max\{y^*a: y^*A \leqslant c\},\end{aligned}$$

because every vector in R_m is the difference of two nonnegative vectors.

Conversely the theorem (with $X = R_n$) can be obtained from either one of the corollaries. To obtain the theorem from Corollary 3.13.12 replace in this corollary x by $x_1 - x_2$ with $x_1 \geqslant 0$ and $x_2 \geqslant 0$, and to obtain this corollary from Corollary 3.13.15 replace in the latter $AX \geqslant a$ by $Ax - s = a$, $s \geqslant 0$.

We cannot replace in Corollary 3.13.12 $X = R_n$ or $Y = R_m$ by an infinite-dimensional space. We have already seen one counterexample in 3.11 where $\alpha = \inf\{-x: x \leqslant 1/i, i = 1, 2, \ldots\}$ to which we can add the constraint $x \geqslant 0$. Then $\alpha = 0$ and $x_o = 0$ but no y_0^* exists if $Y = l_\infty$, with the norm topology replaced by the weak topology of the dual pair (l_∞, l_1). Dualizing this example we get the problem of finding $\alpha = \inf\{\Sigma\xi_i/i: \Sigma\xi_i \geqslant 1, \xi_i \geqslant 0\}$, with $x = (\xi_1, \xi_2, \ldots) \in l_1$. Now $\alpha = 0$ and $y_0^* = 0$ but no x_o exists.

The most symmetric of the foregoing results is Corollary 3.13.12. Usually linear programming is formulated as it is there. Instead of $y_0^*(Ax_o - a) = 0$ as in the theorem we then have $y_0^*(Ax_o - a) = 0$ as well as $z_0^* x_o = 0$ for some $z_0^* \geqslant 0$ (see the proof of this corollary where $z^* = c - y^*A$). Eliminating z^* we get that $(y_0^*A - c)\, x_o = 0$. The conditions

(3.13.18) $\quad y_0^*(Ax_o - a) = 0 \quad$ and $\quad (y_0^*A - c)\, x_o = 0$

are referred to as *complementary slackness*. The word 'slackness' is related to the so-called *slack variables* $s = Ax - a$ and $z^* = c - y^*A$. Complementary slackness is very important because together with the feasibility of x_o and y_0^* it implies that x_o and y_0^* are optimal solutions.

In 6.1 we shall report the unsuccessful attempt to tackle the linear programming problem by applying a fixed point theorem.

3.14 Fenchel duality

In chapter 2 we considered the problem of finding

$$\alpha = \inf\{f(x): g(x) \leqslant 0, x \in C\} \tag{3.14.1}$$

as well as that of finding

$$\alpha = \inf\{ f(x): x \in G\}, \tag{3.14.2}$$

and in the foregoing three sections we have derived a good many duality theorems for the first of these two, where the inequality was sometimes replaced by an equality or a mixture of both. Now we turn to the second one, and let the bifunction F be defined by

$$F(x,y) = f(x) \quad \text{if} \quad x+y \in G. \tag{3.14.3}$$

From the intuitive reasoning in 2.2 we may expect that under certain conditions

$$\inf_X (f(x) - x_0^* x) + \inf_G x_0^* x = \alpha \tag{3.14.4}$$

for some $x_0^* \in X^*$. The two terms in the left hand side look different, in particular because $x \in X$ in the first and $x \in G$ in the second. By means of the following definition we can make them look more alike.

3.14.5 **Definition.** The *indicator function* $\delta(\cdot : G)$ *of the set* G is defined by

$$\delta(x: G) = 0 \quad \text{if} \quad x \in G$$
$$\delta(x: G) = +\infty \quad \text{if} \quad x \notin G.$$

Then (3.14.2)–(3.14.4) become

$$\alpha = \inf_X (f(x) + \delta(x: G)), \tag{3.14.6}$$

$$F(x,y) = f(x) + \delta(x+y: G), \tag{3.14.7}$$

$$\inf_X (f(x) - x_0^* x) + \inf_X (\delta(x: G) + x_0^* x) = \alpha. \tag{3.14.8}$$

The obvious generalization now is to replace $\delta(x: G)$ by any (convex) function $g(x)$. So let us consider the problem of finding

$$\alpha = \inf_X (f(x) + g(x)), \tag{3.14.9}$$

and let

(3.14.10) $$F(x,y) = f(x)+g(x+y).$$

Then the problem is to formulate assumptions under which

(3.14.11) $$\inf_X (f(x)-x_o^*x)+\inf_X (g(x)+x_o^*x) = \alpha,$$

for some $x_o^* \in X^*$.

3.14.12 **Theorem.** Let X be a locally convex topological vector space with dual X^*, and let f and g be convex functions from X to $(-\infty, +\infty]$, so that f and g must not be $-\infty$ anywhere. If α, defined by (3.14.9), is finite then (3.14.11) holds for some $x_o^* \in X^*$ if the following regularity conditions are satisfied:

(3.14.13) $\operatorname{dom} f \cap \operatorname{int} \operatorname{dom} g \neq \emptyset$ (hence $\operatorname{int} \operatorname{dom} g \neq \emptyset$),
(3.14.14) g is u.s.c. at some point $\hat{x}$ of $\operatorname{dom} f \cap \operatorname{int} \operatorname{dom} f$, if X is infinite-dimensional.

Proof. Let F be as in (3.14.10) and let

$$V = \{(y,\mu): \mu \in R, \mu \geqslant F(x,y) \text{ for some } x \in X\};$$

then

$$V = \{(y,\mu): \mu \geqslant f(x)+g(x+y) \text{ for some } x \in X\}.$$

We claim that $(0, f(\hat{x})+g(\hat{x})+1) \in \operatorname{int} V$. To see this let U be a neighbourhood of the origin of X such that $\hat{x}+U \subset \operatorname{dom} g$, and such that $g(\hat{x}+y) < g(\hat{x})+\frac{1}{2}$ if $y \in U$. This is possible because of the regularity conditions. Now we define a neighbourhood W of $(0, f(\hat{x})+g(\hat{x})+1)$ by

$$W = \{(y,\mu): |\mu - f(\hat{x}) - g(\hat{x}) - 1| < \tfrac{1}{2}, y \in U\}.$$

Then $W \subset V$, for if $(y,\mu) \in W$ then $\mu - f(\hat{x}) - g(\hat{x}) > \frac{1}{2}$ and $g(\hat{x}+y) < g(\hat{x})+\frac{1}{2}$, hence $\mu > f(\hat{x})+g(\hat{x}+y)$, so that $(y,\mu) \in V$.

The indicated point, therefore, lies in the interior of V, which means that we may apply Theorems 3.8.14 and 3.8.11 again, and hence for some

$$\begin{aligned} x_o^* \in X^*, \alpha &= \sup_{X^*} \inf_{x \in X, y \in X} (f(x)+g(x+y)+x^*y) \\ &= \inf_{x \in X, y \in X} (f(x)+g(x+y)+x_o^*y) \\ &= \inf_{x \in X, z \in X} (f(x)+g(z)+x_o^*z - x_o^*x) \\ &= \inf_X (f(x)-x_o^*x)+\inf_X (g(x)+x_o^*x). \end{aligned}$$

If X is finite-dimensional then condition (3.14.14) is automatically satisfied because a convex function is continuous in the interior of its effective domain. The latter is shown in Appendix C.

3.14.15 **Corollary.** If in addition to the assumptions of the theorem it is assumed that x_o is an optimal solution then

$$f(x)-f(x_o) \geqslant x_o^*(x-x_o) \quad \text{for all} \quad x, \tag{3.14.16}$$

$$g(x)-g(x_o) \geqslant -x_o^*(x-x_o) \quad \text{for all} \quad x. \tag{3.14.17}$$

Proof. The proof follows by observing that, for all x and all y

$$f(x)-x_o^*x+g(y)+x_o^*y \geqslant f(x_o)+g(x_o),$$

and then taking either $y = x_o$ or $x = x_o$.

If g is the indicator function of a convex set G such that $\operatorname{dom} f \cap \operatorname{int} G \neq \varnothing$ (implying that $\operatorname{int} G$ must not be empty) then (3.14.14) is automatically satisfied, and instead of (3.14.11) we get (3.14.4). Then (3.14.17) reduces to

$$x_o^*(x-x_o) \geqslant 0 \quad \text{for all} \quad x \in G. \tag{3.14.18}$$

A remark is in order here. Up to now we have demonstrated the duality theorems for special cases by means of the general theory of sections 3.7 and 3.8. An alternative, however, is possible when separation arguments are applied directly, thereby circumventing this general theory. This will be done for Theorem 3.14.20. If it is done for the present theorem we can slightly weaken the assumptions, for it is sufficient to assume that $\hat{x} \in \operatorname{int} \operatorname{dom} g$ rather than $\hat{x} \in \operatorname{dom} f \cap \operatorname{int} \operatorname{dom} g$. The reader who wants to pursue this should try to separate weakly the sets

$$A = \{(x,\mu) \colon \mu \in R, \mu \geqslant f(x), x \in \operatorname{dom} f\}$$

and

$$B = \{(x,\mu) \colon \mu \in R, \mu \leqslant \alpha - g(x), x \in \operatorname{dom} g\},$$

by showing among other things that $(\hat{x}, \alpha - g(\hat{x}) - 1) \in \operatorname{int} B$ and that $A \cap \operatorname{int} B = \varnothing$.

3.14.19 **Example.** Let X be a Banach space, $f(x) = |x|$ and G a convex set with nonempty interior. Then to find α we have to find the minimum distance from the origin of X to G. Assume that this minimum distance is equal to $|x_o|$ for some $x_o \in G$. Then (3.14.16) becomes $|x|-|x_o| \geqslant x_o^*(x-x_o)$ for some x_o^* and all x. If we take $x=0$ and $x=2x_o$, this leads to $|x_o^*| = x_o^* x_o$, and since always $x_o^* x_o \leqslant |x_o^*| \cdot |x_o|$ it follows from this that $|x_o^*| \geqslant 1$, at least if $x_o \neq 0$. But since $|x_o^*| = x_o^* x_o$ (3.14.16) can be further reduced to $|x| \geqslant x_o^* x$ for all x, so that $|x| \geqslant |x_o^* x|$ for all x, or $|x_o^* x| \leqslant 1$ if $|x|=1$, and by the definition of $|x_o^*|$ it follows from this that $|x_o^*| \leqslant 1$. We find, therefore, that $|x_o^*|=1$, and that $x_o^* x_o = |x_o^*| \cdot |x_o|$, at least if $x_o \neq 0$.

If $x_o^* x_o = |x_o^*| \cdot |x_o|$ we say that x_o and x_o^* are *aligned.* If $X = R_n$ then x_o and x_o^* are aligned if they are equal up to a positive factor. Then the problem to find x_o^* is equivalent to finding a hyperplane through the origin perpendicular to x_o. This hyperplane has the greatest distance to G of all hyperplanes through the origin, because $\inf_X (f(x) - x_o^* x) + \inf_G x_o^* x$ is equal to the supremum over X^* of $\inf_X (f(x) - x^* x) + \inf_G x^* x$. We have found a classical result, namely that in general finding the minimum distance from a point to a convex set amounts to finding the hyperplane through that point having the maximum distance to that set.

The problem of this example is known as the *minimum norm problem* (see also Example 1.2.1).

Let us now see what can be said if the interior of the effective domain of g is empty. We are then forced to consider relative interiors and restrict X to be a Banach space. In the proof of the next theorem we shall not rely on Theorem 3.8.15, but apply a (weak) separation argument directly.

3.14.20 **Theorem.** Let X be a Banach space with dual X^*, and let f and g be convex functions from X to $(-\infty, +\infty]$. If

$$\alpha = \inf_X (f(x) + g(x)) \tag{3.14.21}$$

is finite, then for some $x_o^* \in X^*$

$$\inf_X (f(x) - x_o^* x) + \inf_X (g(x) + x_o^* x) = \alpha \tag{3.14.22}$$

if the following conditions are satisfied:

(3.14.23) $\operatorname{ri}\operatorname{dom} f \cap \operatorname{ri}\operatorname{dom} g \neq \emptyset$,

(3.14.24) f is u.s.c. relative to $L(\operatorname{dom} f)$ at some point $\hat{x}$ of $\operatorname{ri}\operatorname{dom} f$,

(3.14.25) g is u.s.c. relative to $L(\operatorname{dom} g)$ at some point of $\operatorname{ri}\operatorname{dom} g$, and

(3.14.26) $L(\operatorname{dom} f)+L(\operatorname{dom} g)$ is closed.

Proof. The proof is by *weak separation*. Define the sets P, Q and T by

$$P = \{(y,\mu)\colon \mu \in R, \mu \geqslant f(y), y \in \operatorname{dom} f\},$$

$$Q = \{(y,\mu)\colon \mu \in R, \mu \leqslant \alpha - g(y), y \in \operatorname{dom} g\},$$

and
$$T = \{(y,\mu)\colon \mu \in R, y = 0\}.$$

Since α is finite, P is not empty, so that any closed subspace containing P must contain $L(\operatorname{dom} f)$ as well as T, from which it follows that $L(P) = L(\operatorname{dom} f) \times R$. Similarly it can be shown that $L(Q) = L(\operatorname{dom} g) \times R$, so that

$$L(P)+L(Q) = (L(\operatorname{dom} f)+L(\operatorname{dom} g)) \times R.$$

Since $L(\operatorname{dom} f)+L(\operatorname{dom} g)$ is assumed to be closed, $L(P)+L(Q)$ is closed as well.

Further we have that $\operatorname{ri} P \neq \emptyset$, because $(\hat{x}, f(\hat{x})+1) \in \operatorname{ri} P$. For let U be a neighbourhood of the origin of X such that $\hat{x}+U \cap L(\operatorname{dom} f) \subset \operatorname{dom} f$ and such that

$$f(\hat{x}+y) < f(\hat{x})+\tfrac{1}{2} \quad \text{if} \quad y \in U \cap L(\operatorname{dom} f).$$

This is possible because of (3.14.24). Let

$$W = \{(y,\mu)\colon |\mu - f(\hat{x}) - 1| < \tfrac{1}{2}, y \in U\};$$

then $(\hat{x}, 0)+W \cap L(P) \subset P$, as is easily verified. In a similar way it follows that $\operatorname{ri} Q \neq \emptyset$.

Finally we have that $\operatorname{ri} P \cap \operatorname{ri} Q = \emptyset$. For if $(y,\mu) \in P \cap Q$ then, by the definition of α, $\mu = f(y) = \alpha - g(y)$, so that $(y, \mu-\epsilon) \notin P$ for all $\epsilon > 0$, which means that (y,μ) is a boundary point of P relative to $L(P)$.

To sum up, we have that $\operatorname{ri} P \neq \emptyset$, $\operatorname{ri} Q \neq \emptyset$, $\operatorname{ri} P \cap \operatorname{ri} Q = \emptyset$, and $L(P)+L(Q)$ is closed. Moreover, since $\operatorname{dom} f \cap \operatorname{dom} g \neq \emptyset$, $p = (x', f(x')) \in P$ and $q = (x', \alpha - g(x')) \in Q$ for some x', so

$$p-q = (0, f(x')+g(x')-\alpha) \in T \subset L(P)+L(Q).$$

In view of Theorem 3.3.6 and part (B) of its proof we conclude that P and Q can be separated weakly in the space $L(P)+L(Q)$. This means that, for some z_0^* in the dual of $L(\operatorname{dom} f)+L(\operatorname{dom} g)$ and some $\rho \in R$ such that $(z_0^*, \rho) \neq (0,0)$, $\rho\mu - z_0^* x \geqslant \rho\nu - z_0^*$ if $\mu \geqslant f(x)$ for some $x \in \operatorname{dom} f$ and if $\nu \leqslant \alpha - g(y)$ for some $y \in \operatorname{dom} g$. Here ρ cannot be negative because it it were we could make the left-hand side arbitrarily small and the right-hand side arbitrarily large. Also $\rho = 0$ is impossible, because then $z_0^* \neq 0$ and $-z_0^* x \geqslant -z_0^* y$ for all $x \in \operatorname{dom} f$ and all $y \in \operatorname{dom} g$, which means that $\operatorname{dom} f$ and $\operatorname{dom} g$ could be separated weakly in

$$L(\operatorname{dom} f)+L(\operatorname{dom} g).$$

But because of (3.14.23)

$$x' + U \cap L(\operatorname{dom} f) \subset \operatorname{dom} f \quad \text{and} \quad x' + U \cap L(\operatorname{dom} g) \subset \operatorname{dom} g$$

for some x' and some neighbourhood U of the origin of X, so that $-z_0^*(x-y) \geqslant 0$ if $x \in U \cap L(\operatorname{dom} f)$ and $y \in U \cap L(\operatorname{dom} g)$. Given any $q \in L(\operatorname{dom} f)+L(\operatorname{dom} g)$, we have $\lambda q = x - y$ for some $x \in U \cap L(\operatorname{dom} f)$ and some $y \in L(\operatorname{dom} g)$ *if* λ is positive and small enough. Therefore $-z_0^* q \geqslant 0$ for all $q \in L(\operatorname{dom} f)+L(\operatorname{dom} g)$, hence $z_0^* q = 0$ for all q. But since z_0^* is an element of the dual of $L(\operatorname{dom} f)+L(\operatorname{dom} g)$, we would have that $z_0^* = 0$, which contradicts $z_0^* \neq 0$. It follows that $\rho > 0$. Taking $\rho = 1$ and extending the functional $(z_0^*, 1)$ to a functional $(x_0^*, 1)$, with $x_0^* \in X^*$, such that $\mu - x_0^* x \geqslant \nu - x_0^* y$ if $\mu \geqslant f(x)$ for some $x \in \operatorname{dom} f$, and if $\nu \leqslant \alpha - g(y)$ for some $y \in \operatorname{dom} g$, we finally find that $f(x) - x_0^* x \geqslant \alpha - g(y) - x_0^* y$ for all $x \in \operatorname{dom} f$ and all $y \in \operatorname{dom} g$, from which (3.14.22) immediately follows.

The analogue of Corollary 3.14.15 is obvious and is not stated explicitly here.

The following counterexample shows that the condition that $L(\operatorname{dom} f)+L(\operatorname{dom} g)$ is closed may not be omitted.

3.14.27 **Counterexample.** Let $X = l_2$, the space of all sequences $x = (\xi_1, \xi_2, \ldots)$ such that $|x|^2 = \Sigma \xi_i^2$ exists. Let $\operatorname{dom} f = \{x: \xi_1 + \xi_2 = \xi_3 + \xi_4 = \ldots = 0\}$ and $f(x) = 0$ if $x \in \operatorname{dom} f$. Further let $\operatorname{dom} g = \{x: \xi_2 + \xi_3 = \xi_4 + \xi_5 = \ldots = 0\}$ and $g(x) = \xi_1$ if $x \in \operatorname{dom} g$. Then $\operatorname{dom} f \cap \operatorname{dom} g = \{0\}$, and because $\operatorname{dom} f$ and $\operatorname{dom} g$ are closed linear subspaces, $\operatorname{ri} \operatorname{dom} f = \operatorname{dom} f$ and $\operatorname{ri} \operatorname{dom} g = \operatorname{dom} g$, so that (3.14.23) holds. Moreover f and g are continuous on their effective domains. But $\alpha = 0$ and $\inf_X (f(x) - x^* x) + \inf_x (g(x) + x^* x) = -\infty$ for all x^*. This does not agree with the theorem and the only reason can be that $L(\operatorname{dom} f) + L(\operatorname{dom} g)$ is not closed. This can be verified directly. Let

$$x_i = (\xi_{i1}, -\xi_{i1}, \xi_{i2}, -\xi_{i2}, \ldots) \in \operatorname{dom} f$$

and

$$y_i = (-\eta_{i0}, \eta_{i1}, -\eta_{i1}, \eta_{i2}, -\eta_{i2}, \ldots) \in \operatorname{dom} g,$$

and take

$$\xi_{in} = (2n)^{-\frac{1}{2} - 1/i} \quad \text{and} \quad \eta_{in} = (2n+1)^{-\frac{1}{2} - 1/i}.$$

Then $z_i = x_i + y_i$ converges to

$$z = (2^{-\frac{1}{2}} - 1, 3^{-\frac{1}{2}} - 2^{-\frac{1}{2}}, 4^{-\frac{1}{2}} - 3^{-\frac{1}{2}}, \ldots)$$

but there is no $(x, y) \in \operatorname{dom} f \times \operatorname{dom} g$ such that $z = x + y$.

We started this section with a function f and a set G and then transformed this to two functions f and g which play symmetric parts in Theorem 3.14.20, and also in Theorem 3.14.12 if we replace (3.14.13) by the stronger condition $\operatorname{int} \operatorname{dom} f \cap \operatorname{int} \operatorname{dom} g \neq \varnothing$ and require that f and g are both u.s.c. in their effective domains. In the result f is combined with the term $-x_0^* x$ and g with $+x_0^* x$. In other words f and g are combined with linear functionals that add up to zero. Why not then consider the problem of finding

(3.14.28) $$\alpha = \inf_X (f_1(x) + \ldots + f_k(x))$$

and try to formulate conditions under which for some $x_1^*, \ldots, x_k^*$

(3.14.29)
$$\inf_X (f_1(x) + x_1^* x) + \ldots + \inf_X (f_k(x) + x_k^* x) = \alpha, x_1^* + \ldots + x_k^* = 0.$$

We shall restrict outselves to a case similar to that of Theorem 3.14.12.

3.14.30 **Theorem.** Let X be a locally convex topological vector space with dual X^*, and let $f_1,\dots,f_k$ be convex functions from X to $(-\infty,+\infty]$. Then if α, defined by (3.14.28), is finite, (3.14.29) holds for some $x_1^* \in X^*,\dots,x_k^* \in X^*$, provided that

(3.14.31) $D = \operatorname{int}\operatorname{dom} f_1 \cap \dots \cap \operatorname{int}\operatorname{dom} f_k \neq \emptyset$, and

(3.14.32) f_i is u.s.c. at some point of D, $i = 1,\dots,k$.

Proof. We consider only the case where $k = 3$, leaving the generalization of the (inductive) proof to the reader. So let f_1, f_2 and f_3 be given satisfying the required assumptions. Put $f = f_1 + f_2$ and $g = f_3$. Then

$$\operatorname{dom} f = \operatorname{dom} f_1 \cap \operatorname{dom} f_2 \quad \text{and}$$
$$\operatorname{int}\operatorname{dom} f = \operatorname{int}\operatorname{dom} f_1 \cap \operatorname{int}\operatorname{dom}{}_2,$$

hence by Theorem 3.14.12 we have that

$$\alpha = \inf_X (f(x) - x_0^* x) + \inf_X (g(x) + x_0^* x) \quad \text{for some} \quad x_0^*.$$

Repeating the argument with f, g and α replaced by f', g' and α' defined by $f' = f_1$, $g'(x) = f_2(x) - x_0^* x$ and $\alpha' = \inf_X (f(x) - x_0^* x)$, we obtain that

$$\alpha' = \inf_X (f'(x) - x_{00}^* x) + \inf_X (g'(x) + x_{00}^* x) \quad \text{for some} \quad x_{00}^*.$$

Combining everything we get that

$$\alpha = \inf_X (f_1(x) - x_{00}^* x)$$
$$+ \inf_X (f_2(x) - x_0^* x + x_{00}^* x) + \inf_X (f_3(x) + x_0^* x).$$

Notice that in (3.14.31) we may replace $\operatorname{int}\operatorname{dom} f_1$ by $\operatorname{dom} f_1$ and in (3.14.32) we may take $i = 2,\dots,k$ instead of $i = 1,\dots,k$. Also notice that the proof can be given by letting $x = (x_1,\dots,x_n)$, $f(x) = f_1(x_1)$ if $x_1 = x_2 = \dots = x_n$, $f(x) = +\infty$ otherwise, and $g(x) = f_2(x_2) + \dots + f_n(x_n)$.

3.15 Fenchel duality extended

The beauty of Fenchel duality when applied to the problem of finding $\alpha = \inf\{f(x): x \in G\}$ is that G need not be specified in terms of (in)equalities. Only general assumptions such as convexity and existence of interior points are necessary. The multiplier $x_\circ^*$, however, is an element of X^*, and in this respect Lagrangian duality seems to be more general (but see 3.16). If we want the multiplier to be an element of Y^*, with Y not necessarily equal to X, we might consider the bifunction $F(x, y) = f(x)+g(y)$, but it is quickly seen that nothing interesting comes out of this, for $F(x, 0) = f(x)+g(0)$. What we should do is make the argument of g depend on x as well as y. Since this argument must lie in Y it would be necessary to transform x into some y. The simplest way to do this is by introducing a continuous linear mapping $A: X \to Y$, and by letting

$$F(x, y) = f(x)+g(Ax+y). \tag{3.15.1}$$

The corresponding optimization problem is that of finding

$$\alpha = \inf_X (f(x)+g(Ax)), \tag{3.15.2}$$

and the dual problem is that of finding

$$\begin{aligned}\beta &= \sup_{Y^*} \inf_{X,Y} (f(x)+g(Ax+y)+y^*y)\\ &= \sup_{Y^*} \inf_{X,Y} (f(x)+g(y)+y^*y-y^*(Ax))\\ &= \sup_{Y^*} [\inf_X (f(x)-y^*(Ax))+\inf_Y (g(y)+y^*y)],\end{aligned}$$

so that we should be interested in showing the existence of a $y_\circ^* \in Y^*$ for which

$$\inf_X (f(x)-y_\circ^*(Ax))+\inf_Y (g(y)+y_\circ^*y) = \alpha. \tag{3.15.3}$$

If we define $y_\circ^*A$ by $(y_\circ^*A)\,x = y_\circ^*(Ax)$ then $y_\circ^*A \in X^*$ and we can equally well write $y_\circ^*(Ax) = (y_\circ^*A)\,x = y_\circ^*Ax$.

3.15.4 **Theorem.** Let X and Y be locally convex topological vector spaces with duals X^* and Y^*, respectively, A a continuous linear mapping from X to Y, f a convex function from X, and g a convex function from Y, both to $(-\infty, +\infty]$. If α, defined by (3.15.2), is finite then (3.15.3) holds for some $y_\circ^* \in Y^*$, provided that for some $\hat{x} \in X$

(3.15.5) $\hat{x} \in \operatorname{dom} f$, $A\hat{x} \in \operatorname{int} \operatorname{dom} g$, and g is u.s.c. at $\hat{y} = A\hat{x}$.

Proof. The proof is almost identical to that of Theorem 3.14.12. In the present case $V = \{(y, \mu): \mu \geqslant f(x) + g(Ax + y)$ for some $x \in X\}$ and $(0, f(\hat{x}) + g(\hat{y}) + 1) \in \operatorname{int} V$, for there exists a neighbourhood U of the origin of Y such that $\hat{y} + U \subset \operatorname{dom} g$ and $g(\hat{y} + y) < g(\hat{y}) + \frac{1}{2}$ if $y \in U$. Letting

$$W = \{(y, \mu): |\mu - f(\hat{x}) - g(\hat{y}) - 1| < \tfrac{1}{2}, y \in U\},$$

we have $W \subset V$, so that indeed $(0, f(\hat{x}) + g(\hat{y}) + 1) \in \operatorname{int} V$. The rest of the proof then proceeds as before.

3.15.6 **Example.** Consider the problem of finding

$$\alpha = \inf\{f(x): Ax \geqslant a, x \in C\},$$

where x is in a locally convex topological vector space X, C is a convex set of X, f is a convex function from C to $(-\infty, +\infty)$, a is in a locally convex topological vector space Y with positive cone P, and $A: X \to Y$ is continuous and linear. Let

$$F(x, y) = f(x) + \delta(x: C) + \delta(Ax + y: a + P) \tag{3.15.7}$$

then we can apply the theorem if for some $\hat{x}$, $\hat{x} \in C$, $A\hat{x} \in a + \operatorname{int} P$. For replace f in the theorem by $f + \delta(\cdot : C)$ and let $g(y) = \delta(y: a + P)$; then g is u.s.c. at $\hat{y} = A\hat{x}$. The theorem now tells us that, for some $y_0^* \in Y^*$,

$$\inf_C (f(x) - y_0^* Ax) + \inf_{y \geqslant a} y_0^* y = \alpha.$$

The second infimum here would be $-\infty$ if $y_0^* \not\geqslant 0$, hence $y_0^* \geqslant 0$, but then this infimum is attained at $y = a$. If, in addition to the assumptions of this example, we let $f(x) = cx$ with $c \in X^*$ and $C = \{x: x \geqslant 0\}$ (assuming that X has a positive cone as well) then the first infimum would be $-\infty$ if $c - y_0^* A \not\geqslant 0$, hence $y_0^* A \leqslant c$ and the first infimum is zero. This means that the linear optimization problem of finding $\alpha = \inf\{cx: Ax \geqslant a, x \geqslant 0\}$ leads to another linear optimization problem, i.e. that of finding

$$\beta = \sup\{y^* a: y^* A \leqslant c, y^* \geqslant 0\}.$$

Exactly the same result can be found by applying Lagrangian duality.

3.16 On the equivalence of dualities

The contents of this section might be somewhat disappointing to the reader, at least if he has the impression that both Lagrangian duality and Fenchel duality are truly special cases of bifunctional duality. For, except for minor details, Fenchel duality even in its unextended form is just as general as bifunctional duality, and so is Lagrangian duality.

Let us first consider the relation between bifunctional duality and Fenchel duality. So let a bifunction F be given, let $\alpha = \inf_X F(x, 0)$ and let $\beta = \sup_{Y^*} \inf_{X,Y} (F(x, y) + y^*y)$. As before let

$$V = \{(y, \mu) : \mu \geqslant F(x, y) \text{ for some } x\} \quad \text{and} \quad T = \{(y, \mu) : y = 0\}.$$

Further define f' and g' both from $Y \times R$ to R by

$$\text{(3.16.1)} \qquad \operatorname{dom} f' = T, \quad f'(y, \mu) = \mu \quad \text{if} \quad (y, \mu) \in T,$$
$$\text{(3.16.2)} \qquad \operatorname{dom} g' = V, \quad g'(y, \mu) = 0 \quad \text{if} \quad (y, \mu) \in V.$$

Then

$$\inf_{Y \times R} (f'(y, \mu) + g'(y, \mu)) = \inf\{\mu + 0) : \mu \geqslant F(x, 0) \text{ for some } x\} = \inf_X F(x, 0) = \alpha.$$

And

$$\inf_{Y \times R} (f'(y, \mu) - y^*y - \mu^*\mu) + \inf_{Y \times R} (g'(y, \mu) + y^*y + \mu^*\mu)$$

with $y^* \in Y^*$ and $\mu^* \in R$ becomes

$$\inf_R (1 - \mu^*)\mu + \inf_{Y \times R} \{\mu^*\mu + y^*y : \mu \geqslant F(x, y) \text{ for some } x\}.$$

Now we assume that $F(x, y) < +\infty$ for at least one pair (x, y). Then the second infimum is not $+\infty$. If $\mu^* \neq 1$, then the first infimum would be $-\infty$, so that the sum of both terms would also be $-\infty$. When computing the supremum of this sum over all (y^*, μ^*) we can therefore take $\mu^* = 1$. Then the first infimum is zero and the second one becomes

$$\inf\{\mu + y^*y : \mu \geqslant F(x, y) \text{ for some } x\} = \inf_{X,Y} (F(x, y) + y^*y).$$

The conclusion, therefore, is that $\alpha = \beta$ if and only if

$$\inf_{Y\times R}(f'(y,\mu)+g'(y,\mu)) = \sup_{Y^*\times R^*}[\inf_{Y\times R}(f'(y,\mu)-y^*y-\mu^*\mu) \\ +\inf_{Y\times R}(g'(y,\mu)+y^*y+\mu^*\mu)]$$

at least if F is not $+\infty$ everywhere.

As for the solvability of the primal and the dual problems, assume first that $\alpha = f'(y_0,\mu_0)+g'(y_0,\mu_0)$ for some (y_0,μ_0) and that α is not $+\infty$. Then, since f' and g' are infinite outside their effective domains, $y_0 = 0$ and $\mu_0 \geqslant F(x_0, 0)$ for some x_0. Notice that x_0 does not show up explicitly in the corresponding Fenchel duality relation. It follows that $\alpha = \mu_0 \geqslant F(x_0, 0)$ and hence that $F(x_0, 0) = \inf_X F(x, 0)$. Next assume that, for some (y_0^*, μ_0^*),

$$\beta = \inf_{Y\times R}(f'(y,\mu)-y_0^*y-\mu_0^*\mu)+\inf_{Y\times R}(g'(y,\mu)+y_0^*y+\mu_0^*\mu) \\ = \inf_R(\mu-\mu_0^*\mu)+\inf_{Y\times R}\{y_0^*y+\mu_0^*\mu\colon \mu \geqslant F(x,y) \text{ for some } x\}.$$

When $\beta > -\infty$ then $\mu_0^* = 1$ and $\beta = \inf_{X,Y}(F(x,y)+y_0^*y)$, showing that y_0^* is a dual solution. Hence we have the following theorem.

3.16.3 **Theorem.** Given a bifunction F such that the corresponding infimum is not $+\infty$ (so that $F(x,0) < +\infty$ for at least one x) and such that the corresponding supremum is not $-\infty$, then duality with respect to F (including the solvability of both the primal and the dual problems) amounts to Fenchel duality (including the solvability of both the primal and the dual problems) with respect to a suitably chosen problem.

But Fenchel duality in turn can be expressed in terms of Lagrangian duality. Consider the problem of finding $\alpha = \inf_X(f(x)+g(x))$ and $\beta = \sup_{X^*}[\inf_X(f(x)-x^*x)+\inf_X(g(x)+x^*x)]$.

Then

$$\alpha = \inf\{f(x_1)+g(x_2)\colon x_1 = x_2, (x_1,x_2)\in \operatorname{dom} f\times \operatorname{dom} g\}.$$

Letting

$$X' = X\times X, f'(x') = f(x_1)+g(x_2),\ B'x' = x_2-x_1$$

with $x' = (x_1, x_2)$ and $C' = \operatorname{dom} f\times \operatorname{dom} g$, we obtain $\alpha = \inf\{f'(x')\colon B'x' = 0, x'\in C'\}$, to which we may apply Lagrangian duality. If f and g are convex functions then f' is a convex function and C' is a convex set. Clearly B' is a continuous linear function. Assume that $B'(L(C')) = X'$. Now if f and g are u.s.c. at some $\hat{x} \in \operatorname{ri} \operatorname{dom} f \cap \operatorname{ri} \operatorname{dom} g$, then f' is u.s.c. at $(\hat{x}, \hat{x}) \in \operatorname{ri} C'$ and Theorem 3.12.1 applies if X is a Banach space. So if α is finite then for some $x_0^* \in X^*$ we have that

$$\begin{aligned}\alpha = \beta &= \inf_{C'} (f'(x') + x_0^* B'x') \\ &= \inf\{f(x_1) + g(x_2) + x_0^*(x_2 - x_1) \colon x_1 \in \operatorname{dom} f, x_2 \in \operatorname{dom} g\} \\ &= \inf_X (f(x) - x_0^* x) + \inf_X (g(x) + x_0^* x).\end{aligned}$$

3.16.4 **Theorem.** The result of Theorem 3.14.20 can be derived from Lagrangian duality if conditions (3.14.24) and (3.14.25) are replaced by the tighter condition that f and g are u.s.c. at the *same* point, and if the condition that $L(\operatorname{dom} f) + L(\operatorname{dom} g)$ is closed is replaced by the condition that $B'(L(C')) = X'$ with $C' = \operatorname{dom} f \times \operatorname{dom} g$, $X' = X \times X$ and $B'(x_1, x_2) = x_2 - x_1$.

If $\operatorname{int} \operatorname{dom} g \neq \emptyset$ then automatically $B'(L(C')) = X'$. This fact leads to the next theorem.

3.16.5 **Theorem.** The result of Theorem 3.14.12 can be derived from the Lagrangian duality if condition (3.14.14) is replaced by the condition that *both* f and g are u.s.c. at some point $\hat{x} \in \operatorname{int} \operatorname{dom} f \cap \operatorname{int} \operatorname{dom} g$, which implies that condition (3.14.13) must be replaced by $\operatorname{int} \operatorname{dom} f \cap \operatorname{int} \operatorname{dom} g \neq \emptyset$.

Clearly if, for some $x'_0 \in X'$, $f'(x'_0) = \inf\{f'(x') \colon B'x' = 0, x' \in C'\}$ and $B'x'_0 = 0$, $x'_0 \in C'$, then $x'_0 = (x_0, x_0)$ for some x_0 and $\alpha = f(x_0) + g(x_0)$. Hence the primal solvability of the Fenchel duality problem follows from the primal solvability of the corresponding Lagrangian duality problem.

Broadly speaking then we may say that bifunctional duality, Fenchel duality and Lagrangian duality are equivalent. Yet it is extremely convenient to have available a number of results pertaining to special types of duality, so that after all we should not be too disappointed.

4
A global approach by conjugate duality

4.1 Converse duality; conjugate functions

So far the main line of thinking has been to make certain assumptions about the primal problem in order to be able to derive the equality of the primal infimum and the dual supremum, as well as the existence of a dual solution y_0^*. But since the dual problem can be formulated independently of whether or not certain duality results hold, it is natural to consider the reverse line of thinking as well.

This makes it necessary to consider the dual bifunction F^d defined by

$$F^d(y^*, x^*) = \inf_{X,Y} (F(x, y) - x^*x + y^*y) \tag{4.1.1}$$

and the dual perturbation function p^d defined by

$$p^d(x^*) = \sup_{Y^*} F^d(y^*, x^*). \tag{4.1.2}$$

Consequently we must assume that X, as well as Y, is topologized, so that the dual X^* of X, as well as the dual Y^* of Y, is properly defined. Thus our starting point would be the problem of finding

$$\beta = \sup_{Y^*} F^d(y^*, 0). \tag{4.1.3}$$

If we want to construct the 'dual' of this problem we must introduce bilinear forms $x^{**}x^*$ and $y^{**}y^*$ with $x^{**} \in X^{**}$ and $y^{**} \in Y^{**}$, so that the 'dual' decision space is X^{**} which in general cannot be identified with X, and the 'dual' perturbation space is Y^{**} which in general cannot be identified with Y. Hence in general there is not much hope of getting back to the original problem, i.e. the problem of finding $\alpha = \inf_X F(x, 0)$ by dualizing the dual problem of finding β. Yet this is what we want, so we shall assume that (X, X^*) and (Y, Y^*) are two pairs of dual spaces, even though in certain cases there may be an alternative. To give some idea of the latter, assume that X and X^* form a dual pair, but that Y is equal to l_∞ complete with its norm topology. Then l_∞^* cannot be identified with l_1 (which would be possible if we replaced the norm topology of l_∞ by say the weak topology of the dual pair (l_∞, l_1)). On the other hand the interior of the nonpositive orthant of $Y = l_\infty$ is not empty (and this interior would be empty if we considered the weak topology) which, as we have seen is an important fact in certain duality theorems. Now suppose that for a given problem it can be shown by the very nature of that problem that $y_0^* \in l_\infty^*$ can be represented by an element of l_1. Then we could write $\beta = \sup_{l_1} F^d(y^*, 0)$ instead of $\beta = \sup_{l_\infty^*} F^d(y^*, 0)$, and constructing the dual of the dual would lead to $Y^{**} = l_1^* = l_\infty$, so that we would be back to where we started. A similar way out might be possible with respect to X instead of Y.

These considerations lead to the following definitions.

4.1.4 **Definition.** Given a bifunction F from $X \times Y$ to $[-\infty, +\infty]$, the *bidual* (or *biadjoint*) *bifunction* F^{dd} is defined by

$$F^{dd}(x, y) = \sup_{Y^*, X^*} (F^d(y^*, x^*) + x^*x - y^*y), \tag{4.1.5}$$

where F^d is defined by (4.1.1), and the *bidual* (or *biadjoint*) *perturbation function* p^{dd} is defined by

$$p^{dd}(y) = \inf_X F^{dd}(x, y). \tag{4.1.6}$$

The reason why we take sup in (4.1.5) and inf in (4.1.1) is that, if F is convex, F^{dd} is convex too, but F^d is concave, so that $-F^{dd}$ is concave and $-F^d$ is convex. Indeed we have from (4.1.5) that

$$-F^{dd}(-x, -y) = \inf_{Y^*, X^*} (-F^d(y^*, x^*) - y^*y + x^*x)$$

and this is entirely similar to (4.1.1). The latter equality also shows that, strictly speaking, we should say that $-F^{dd}$ with the signs of its arguments reversed is the dual of minus the dual of F, but instead we shall simply say that *F^{dd} is the dual of the dual of F*, tacitly assuming that the dual of a concave problem is constructed slightly differently from that of a convex problem.

Unfortunately, notwithstanding the fact that we restricted ourselves to dual pairs of spaces, we have not yet reached our aim of perfect symmetry, for in general $F^{dd} \neq F$. Equality is obtained here, however, if we assume that $F(x, y) > -\infty$ for all (x, y), so that

$$\sup_{Y^*, X^*} (F(x', y') - x^*x' + y^*y' + x^*x - y^*y) = +\infty$$

unless $x' = x$ and $y' = y$, and that we may interchange inf and sup in the following identity, which is implied by (4.1.5) and (4.1.1),

(4.1.7)
$$F^{dd}(x, y) = \sup_{Y^*, X^*} \inf_{X, Y} (F(x', y') - x^*x' + y^*y' + x^*x - y^*y).$$

For after interchanging sup and inf here we may put $x' = x$ and $y' = y$ so that the right-hand side reduces to $F(x, y)$.

Obviously, $F^{dd} = F$ if and only if

(4.1.8)
$$F(x, y) = \sup_{Y^*, X^*} \inf_{X, Y} (F(x', y') - x^*x' + y^*y' + x^*x - y^*y).$$

Interestingly enough, in (3.10) we came across a similar situation. There we found results in terms of the Lagrangian which are symmetric with respect to the primal and the dual by assuming that

(4.1.9)
$$F(x, y) = \sup_{Y^*} (L(x, y^*) - y^*y) = \sup_{Y^*} \inf_{Y} (F(x, y') + y^*y' - y^*y).$$

The only difference from (4.1.8) is that here $F(x, y)$ is computed by first transforming a function of y into a function of y^* and then transforming the latter into a function of y again, whereas in (4.1.8) a function of (x, y) is transformed into a function of (y^*, x^*) which is then transformed back into a function of (x, y).

As a third example consider our well-known problem of establishing the equality of α and β, which is equivalent to the equality of $p(0)$ and $p^d(0) = \sup_{Y^*} \inf_Y (p(y) + y^*y)$. If $p(y)$ is nowhere $-\infty$ and if sup and inf may be interchanged here then $p(0) = p^d(0)$. Again a function of y is transformed into a function of y^* which is transformed back into a function of y.

All the time we are applying the same type of transformation, and all the time we require that, after applying such a transformation twice, we come out with the function or function value we started from. So independently of any optimization problem let us introduce the following notions.

4.1.10 **Definition.** Let (Z, Z^*) be a dual pair of locally convex topological vector spaces, and h a not necessarily convex or concave function $h: Z \to [-\infty, +\infty]$. Then

$$h^*(z^*) = \sup_Z (z^*z - h(z)) \tag{4.1.11}$$

is called the *conjugate (function) of h*, and

$$h^{**}(z) = \sup_{Z^*} (z^*z - h^*(z^*)) \tag{4.1.12}$$

is called the *biconjugate (function) of h.*

The function h^* is also termed the *Young transform* or *Fenchel transform* of h. Notice that the definitions would also apply if Z were not a member of a dual pair. All that is needed is that Z has a dual.

The conjugate of a convex function is not concave as might perhaps have been expected after the discussion leading to Definition 4.1.10. In fact h^* is a convex function even if h is not. This is easily shown.

A geometric interpretation of the conjugate is obtained if we consider the conjugate of the indicator function $\delta(\cdot : G)$ of some subset G of Z. In particular let $G = \operatorname{epi} f \subset Z = X \times R$ of a function $f: X \to [-\infty, +\infty]$. Then

$$\begin{aligned} f^*(x^*) &= \sup_X (x^*x - f(x)) = \sup_{X,Y} \{x^*x - \mu : \mu \geqslant f(x)\} \\ &= \sup_Z \{(x^*, -1) z : z \in \operatorname{epi} f\} = \delta^*((x^*, -1) : \operatorname{epi} f). \end{aligned}$$

Hence $f^*(x^*)$ is equal to the value of the conjugate of the indicator function of epi f at the point $(x^*, -1)$. We may also write

$$f^*(x^*) = \inf\{\gamma : \gamma \geqslant x^*x - \mu \text{ for all } z = (x, \mu) \in \operatorname{epi} f\}.$$

But $(0, -\gamma)$ is the intercept of the μ-axis with the hyperplane $H(\gamma) = \{(x, \mu) : x^*x - \mu = \gamma\}$, so that $(0, -f^*(x^*))$ is the highest intercept of the μ-axis with hyperplanes $H(\gamma)$ which have epi f at one side, and which may equally well be taken as hyperplanes *supporting* epi f. This at the same time justifies the term *support function of G* for $\delta^*(\cdot : G)$.

The geometric interpretation of $f^*(x^*)$ makes it immediately clear why the functions f_1, f_2 and f_3 of Figure 2 all have the same conjugate. Clearly f_1 is not convex and the figure gives the right impression that

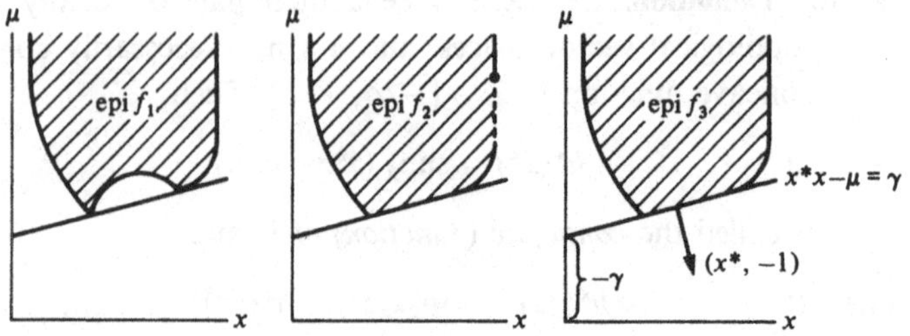

Figure 2. Effects of conjugation.

conjugation ignores 'nonconvexities'. But f_2 is nicely convex. So we see that conjugation ignores something else as well and that is the fact that f_2 is not everywhere lower semi-continuous (l.s.c.), whereas f_3 is. It is this fact that causes trouble when trying to prove that inf = sup for a problem where everything is convex. We have already seen an instance of this in Example 3.5.9, where

$$p(y) = -y^{\frac{1}{2}} \quad \text{if} \quad y > 0,\ p(0) = 1 \quad \text{and} \quad p(y) = +\infty \quad \text{if} \quad y < 0.$$

Like h^*, h^{**} is a convex function, so if we want $h^{**} = h$ then the first requirement must be that h is convex. Figure 2 shows that this is not a sufficient condition, for $f_2^{**} = f_3^{**}$ and this must be different from at least one of f_2 and f_3 (in fact it only differs from f_2). That it is quite desirable to have equality of h^{**} and h can be seen by fixing x, letting $Z = Y$ and $h(y) = F(x, y)$. Then

$$\begin{aligned} h^{**}(y) &= \sup_{Y^*}(y^*y - \sup_Y (y^*y' - F(x, y'))) \\ &= \sup_{Y^*} \inf_Y (F(x, y' - y^*y' + y^*y) \end{aligned}$$

and if we replace y^* by $-y^*$ this becomes

$$\sup_{Y^*} \inf_Y (F(x, y') + y^*y' - y^*y);$$

hence if $h^{**} = h$ then F can be recovered from L by (4.1.9).

Similarly, if we let $Z = X \times Y$ and $h(x, y) = F(x, y)$ then we have that $h^{**} = h$ is equivalent to $F^{dd} = F$.

And if we let $Z = Y$ and $h(y) = p(y)$ then $p(0) = p^d(0)$ if and only if $h^{**}(0) = h(0)$, so that h need only be reproduced at $y = 0$. In Example 3.5.9 this does not happen because $p^d(0) = 0$ and $p(0) = 1$.

It is easily shown that

$$(4.1.13) \qquad q^*(y^*) \triangleq -L(x, -y^*) = \sup_Y (y^*y - F(x, y))$$

hence, for fixed x, q^* is the conjugate of $F(x, \cdot)$. When

$$F(x, y) = \sup_{Y^*} (L(x, y^*) - y^*y) = \sup_{Y^*} (y^*y - (-L(x, -y^*)))$$

we have that $F(x, \cdot)$ is the conjugate of q^*, and hence that $F(x, \cdot)$ is the biconjugate of itself. Conversely, if $F^{**}(x, \cdot) = F(x, \cdot)$ with conjugations taken with respect to y alone, then $F(x, \cdot)$ is the conjugate of q^*.

Further we have that

$$(4.1.14) \qquad -F^{\mathrm{d}}(-y^*, x^*) = \sup_{X, Y} (x^*x + y^*y - F(x, y))$$

and that

$$(4.1.15) \qquad F^{\mathrm{dd}}(x, y) = F^{**}(x, y)$$

With conjugations taken with respect to both x and y.

Finally we have that

$$(4.1.16) \qquad p^{\mathrm{d}}(0) = p^{**}(0).$$

Conjugations are also useful when formulating Fenchel duality. Instead of writing, as in (3.14.22),

$$\inf_X (f(x) - x_0^*x) + \inf_X (g(x) + x_0^*x) = \inf_X (f(x) + g(x))$$

we can now write

$$(4.1.17) \qquad -f^*(x_0) - g^*(-x_0^*) = \inf_X (f(x) + g(x)).$$

Notice that in general $(-f)^* \neq -f^*$, for $(-f)^*$ is convex but $-f^*$ is concave. In a case like (4.1.17) it is convenient to introduce the *concave conjugate* of a function $h: Z \to [-\infty, +\infty]$, given by $h_*(z^*) = \inf_Z (z^*z - h(z))$. It is easily verified that $h_*(z^*) = -(-h)^*(-z^*)$, and that $h_{**} = -(-h)^{**}$. If in (4.1.17) we replace g by $-g$ then we get

$$(4.1.18) \qquad g_*(x_0^*) - f^*(x_0^*) = \inf_X (f(x) - g(x))$$

and we have $\alpha = \beta$ if and only if

$$(4.1.19) \quad \inf_X (f(x) - g(x)) = \sup_{X^*} (g_*(x^*) - f^*(x^*)).$$

In the relevant Fenchel duality theorem f is still convex, but g must be concave. In the literature g_* is often replaced by g^* and one should know from the context which conjugation is meant.

4.2 Closed functions

Let us look at Figure 2 again. We have remarked already that the reason why $f_2^* = f_3^*$ although $f_2 \neq f_3$ is due to the fact that f_2 is not everywhere l.s.c. Another way of putting this is to say that the *set* $\mathrm{epi}\, f_2$ is not closed, and this suggests that we introduce the idea of closed *functions*.

4.2.1 **Definition.** If Z is a locally convex topological vector space, then $h: Z \to [-\infty, +\infty]$ is called *closed* if $\mathrm{epi}\, h$ is closed in $Z \times R$. And the *closure of* h, denoted by $\mathrm{cl}\, h$, is defined by

$$(\mathrm{cl}\, h)(z) = \inf\{\nu : \nu \in R, (z, \nu) \in \mathrm{cl}\, \mathrm{epi}\, h\}.$$

Various definitions appear in the literature. The present one is taken from Ioffe & Tikhomirov (1968). Another comes from Rockafellar (1970*b*):

(4.2.2)

$$\begin{array}{llll} \mathrm{cl}_R h = \mathrm{cl}\, h, & & \text{if } (\mathrm{cl}\, h)(z) > -\infty & \text{for } \textit{all } z \in Z. \\ (\mathrm{cl}_R h)(z) = -\infty & \text{for all } z, & \text{if } (\mathrm{cl}\, h)(z) = -\infty & \text{for } \textit{some } z \in Z. \end{array}$$

The following lemma summarizes a number of useful properties relating to closure and conjugation.

4.2.3 **Lemma.**

(*a*) $\mathrm{epi}\, \mathrm{cl}\, h = \mathrm{cl}\, \mathrm{epi}\, h$, hence $\mathrm{cl}\, h$ is a closed function, and $h = \mathrm{cl}\, h$ if and only if h is a closed function.

(*b*) $\mathrm{cl}\, h$ is a convex function if h is a convex function.

(*c*) If $h \leqslant k$, then $\mathrm{cl}\, h \leqslant \mathrm{cl}\, k$, $h^* \geqslant k^*$, $h^{**} \leqslant k^{**}$.

(*d*) h^* and h^{**} are closed convex functions.

(*e*) $h^{**} \leqslant h$.

(*f*) $(\mathrm{cl}\, h)^{**} \leqslant h^{**} \leqslant \mathrm{cl}\, h \leqslant h$.

Proof.

(*a*) Let $(z, \mu) \in \mathrm{epi}\, \mathrm{cl}\, h$. Then $\mu \geqslant (\mathrm{cl}\, h)(z)$, hence for all $\epsilon > 0$ there exists a ν such that $\nu < (\mathrm{cl}\, h)(z) + \epsilon \leqslant \mu + \epsilon$ and such that

$(z, \nu) \in \text{cl epi}\, h$, so that $(z, \mu+\epsilon) \in \text{cl epi}\, h$ and also $(z, \mu) \in \text{cl epi}\, h$. It follows that $\text{epi cl}\, h \subset \text{cl epi}\, h$.

Conversely let $(z, \mu) \in \text{cl epi}\, h$. Then $\mu \geqslant (\text{cl}\, h)(z)$, hence $(z, \mu) \in \text{epi cl}\, h$, so that $\text{cl epi}\, h \subset \text{epi cl}\, h$.

(*b*) Let h be a convex funtion. Then $\text{epi}\, h$ is a convex set, and so is $\text{cl epi}\, h$, so that $\text{cl}\, h$ is a convex function.

(*c*) This is a simple consequence of the definitions.

(*d*) Let every neighbourhood of (z_0^*, μ_0^*) contain a point $(z^*, \mu^*) \in \text{epi}\, h^*$. Then for all $z \in Z$ we have that $\mu^* \geqslant z^*z - h(z)$, hence that $\mu_0^* \geqslant z_0^*z - h(z)$, or $\mu_0^* \geqslant h^*(z_0^*)$ which means that $(z_0^*, \mu_0^*) \in \text{epi}\, h^*$, so that $\text{epi}\, h^*$ is closed. In a similar way it can be shown that h^{**} is closed.

(*e*) By definition

$$h^{**}(z) = \sup_{Z^*} \inf_Z (z^*z - z^*z' + h(z'))$$
$$\leqslant \inf_Z \sup_{Z^*} (z^*z - z^*z' + h(z')),$$

and $$\sup_{Z^*}(z^*z - z^*z' + h(z')) = +\infty$$

if $z' \neq z$ and $h(z') > -\infty$ for all z', so that then $\inf\sup = h(z)$. If $h(z') = -\infty$ for some z', then $h^{**}(z) = -\infty$ for all z.

(*f*) The inequality $\text{cl}\, h \leqslant h$ is an immediate consequence of the definition. From this and part (*c*) it follows that $(\text{cl}\, h)^{**} \leqslant h^{**}$. And by parts (*d*), (*e*) and (*c*) it follows that $h^{**} = \text{cl}\,(h^{**}) \leqslant \text{cl}\, h$.

The next lemma requires a *strong separation* argument.

4.2.4 **Lemma.** Let h be a convex function $h\colon Z \to [-\infty, +\infty]$, If $(\text{cl}\, h)(z')$ is finite for some z', then $(\text{cl}\, h)^{**}$, h^{**}, $\text{cl}\, h$ and h are nowhere $-\infty$.

Proof. Because of the preceding lemma, part (*f*), it is sufficient to show that $(\text{cl}\, h)^{**}$ is nowhere $-\infty$. Put $\mu' = (\text{cl}\, h)(z') - 1$. Then $(z', \mu') \notin \text{epi cl}\, h = \text{cl epi}\, h$, hence we can separate (z', μ') strongly from $\text{epi cl}\, h$, which means that for some $\rho \in R$ and some $z_0^* \in Z^*$

$$\rho\mu - z_0^*z > \rho\mu' - z_0^*z' \quad \text{if} \quad \mu \geqslant (\text{cl}\, h)(z).$$

It follows that $\rho \geqslant 0$, and even that $\rho > 0$, as can be seen by taking $z = z'$ and $\mu = \mu' + 1$. With $\rho = 1$ we obtain that $\inf_Z ((\text{cl}\, h)(z) - z_0^*z)$ is not $-\infty$, so that $(\text{cl}\, h)^*(z_0^*)$ is not $+\infty$ and $(\text{cl}\, h)^{**}$ is nowhere $-\infty$.

4.2.5 **Theorem.** Let h be a closed convex function

$$h: Z \to [-\infty, +\infty].$$

Then $h^{**} = h$, unless $h(z) = -\infty$ for some but not all z.

Proof. The provision in the theorem is necessary, because if $h(z) = -\infty$ for some z, then $h^{**}(z) = -\infty$ for all z. If $h(z) = -\infty$ for all z then, of course, $h^{**} = h$ and this is also true if $h(z) = +\infty$ for all z. So we may now assume that $h(z) > -\infty$ for all z and that $h(z')$ is finite for some z'.

Let any z'' be given. We shall show that $h^{**}(z'') = h(z'')$. Consider the trivial optimization problem to find

$$h(z'') = \alpha = \inf\{h(z): z = z''\}.$$

Let the bifunction be $F(z, y) = h(z)$ if $z = z'' + y$, and $F(z, y) = +\infty$ otherwise. Let

$$V = \{(y, \mu): \mu \geqslant F(z, y) \text{ for some } z\}.$$

Then

$$V = \{(y, \mu: \mu \geqslant h(z'' + y)\} = (-z'', 0) + \operatorname{epi} h.$$

But epi h is closed; hence V is closed, so that by Theorem 3.7.1 F is normal. The supremum of the dual problem is

$$\begin{aligned}\beta &= \sup_{Y^*} \inf_{Z, Y} (F(z, y) + y^* y) = \sup_{Y^*} \inf_Y (h(z'' + y) + y^* y) \\ &= \sup_{Y^*} \inf_Y (h(y) + y^* y - y^* z'') = h^{**}(z'').\end{aligned}$$

Since $h(z')$ is finite it follows from the lemma that $\beta \neq -\infty$; hence $(\alpha, \beta) \neq (+\infty, -\infty)$ and by Theorem 3.6.1 it follows that $h(z'') = \alpha = \beta = h^{**}(z'')$.

4.2.6 **Corollary.** Let h be a convex function $h: Z \to [-\infty, +\infty]$. Then $h^{**} = \operatorname{cl} h$, unless $(\operatorname{cl} h)(z) = -\infty$ for some but not all z.

Proof. By Lemma 4.2.3 part (*b*), cl h is convex, hence by the theorem $(\operatorname{cl} h)^{**} = \operatorname{cl} h$, and by Lemma 4.2.3 part (*f*), $h^{**} = \operatorname{cl} h$, unless $(\operatorname{cl} h)(z) = -\infty$ for some but not all z.

The theorem can also be proved directly by a strong separation argument rather than using normality and Theorem 3.6.1. The ingredients of this alternative proof, however, would be much the same as those of the one given.

The conjecture that the provision in the corollary might be replaced by: unless $h(z) = -\infty$ for some but not all z, would be wrong.

4.2.7 **Counterexample.** Let Z be the space of all finite sequences $z = (\zeta_1, \zeta_2, \ldots, \zeta_n, 0, \ldots)$ and Z^* the space of all finite sequences $z^* = (\zeta_1^*, \zeta_2^*, \ldots, \zeta_m^*, 0, \ldots)$. Let $z^*z = \Sigma\zeta_i^*\zeta_i$ and consider the σ-topology defined by this bilinear form. Then σ-convergence is simply convergence component-wise. Let $h(z) = \Sigma\zeta_i$ if $z \leqslant 0$ (which means that $\zeta_i \leqslant 0$ for all i), and $h(z) = +\infty$ otherwise. Then $\operatorname{dom} h = \{z: z \leqslant 0\}$ is closed, so that $(\operatorname{cl} h)(z) = +\infty$ if $z \nleqslant 0$.

And $(\operatorname{cl} h)(z) = -\infty$ if $z \leqslant 0$. To see this let $z = (\zeta_1, \ldots, \zeta_n, 0, \ldots) \leqslant 0$ be given. Let $\lambda > 0$ and define z_k by

$$z_k = \{\zeta_1, \ldots, \zeta_n, -\lambda/k, \ldots, -\lambda/k, 0, \ldots),$$

where the component $-\lambda/k$ is repeated k times. Further let $\mu_k = h(z_k) = h(z) - \lambda$. Then $(z_k, \mu_k) \in \operatorname{epi} h$ and (z_k, μ_k) converges to $(z, h(z) - \lambda)$ if k tends to infinity, so that $(z, h(z) - \lambda) \in \operatorname{cl} \operatorname{epi} h$ for all positive λ, which means that $(\operatorname{cl} h)(z) = -\infty$.

Let us now compute $h^*(z^*)$. Given any $z^* = (\zeta_1^*, \ldots, \zeta_m^*, 0, \ldots)$, take $z \leqslant 0$ such that $\zeta_i = 0$ if $i \neq m+1$ and such that $\zeta_{m+1} = -\omega$. Then

$$h^*(z^*) = \sup\{\Sigma(\zeta_i^* - 1)\zeta_i : \zeta_i \leqslant 0\} \geqslant \omega \text{ for all } \omega > 0;$$

hence $h^*(z^*) = +\infty$ for all $z^* \in Z^*$ and so $h^{**}(z) = -\infty$ for all $z \in Z$. It follows that $h^{**}(z) \neq (\operatorname{cl} h)(z)$ if $z \nleqslant 0$.

In this counterexample $\operatorname{ri} \operatorname{dom} h$ is empty. The next theorem takes care of this.

4.2.8 **Theorem.** Let h be a convex function $h: Z \to (-\infty, +\infty]$; hence $h(z) > -\infty$ for all z. Let $\operatorname{ri} \operatorname{dom} h \neq \varnothing$ and let h be u.s.c. at all $z \in \operatorname{ri} \operatorname{dom} h$. Then $h^{**} = \operatorname{cl} h$ and $(\operatorname{cl} h)(z) = h(z)$ if $z \in \operatorname{ri} \operatorname{dom} h$.

Proof. We follow a line of reasoning similar to that in the proof of Theorem 4.2.5. So let z'' be any element of $\operatorname{ri} \operatorname{dom} h$ and consider the problem of finding $h(z'') = \alpha = \inf\{h(z): z = z''\}$. Then α is finite and again $V = (-z'', 0) + \text{epi}\, h$. Moreover $T \cap \text{ri}\, V \neq \emptyset$ with $T = \{(z, \mu): z = 0\}$. To see this let U be a neighbourhood of the origin such that

$$z'' + U \cap L(\text{dom}\, h) \subset \text{dom}\, h$$

and $$h(z) < h(z'') + \tfrac{1}{2} \quad \text{if} \quad z \in z'' + U \cap L(\text{dom}\, h).$$

Further let

$$W = \{(z, \mu): |\mu - h(z'') - 1| < \tfrac{1}{2}, z \in z'' + U\}.$$

Then W is a neighbourhood of $(z'', h(z'') + 1) \in \text{epi}\, h$. Since $L(\text{epi}\, h) = L(\text{dom}\, h) \times R$ it is now easily verified that $W \cap ((z'', 0) + L(\text{epi}\, h)) \subset \text{epi}\, h$, which means that $(z'', h(z'') + 1) \in \text{ri}\,\text{epi}\, h$ and that $(0, h(z'') + 1) \in \text{ri}\, V$. Normality now follows from Theorem 3.7.2

By Theorem 3.6.1 we have again that $h(z'') = \alpha = \beta = h^{**}(z'')$ and by Lemma 4.2.3 part (*f*) this is equal to $(\text{cl}\, h)\,(z'')$. Since then $(\text{cl}\, h)(z'')$ is finite, by Lemma 4.2.4. $\text{cl}\, h$ is nowhere $-\infty$, and by Corollary 4.2.6 $h^{**} = \text{cl}\, h$.

Notice that h is not automatically u.s.c. in $\text{ri}\,\text{dom}\, h$, not even if h is linear! For take h as in the counterexample, except that $\text{dom}\, h = Z$. Let $z_k = (1/k, \ldots, 1/k, 0, \ldots)$ with the component $1/k$ repeated k times. Then z_k tends to 0 if k tends to infinity, but $h(z_k) = 1$ and $h(0) = 0$.

We close this section by relating h^{**} to $\text{cl}_R h$.

4.2.9 **Theorem.** Let h be a convex function $h: Z \to [-\infty, +\infty]$. Then $h^{**} = \text{cl}_R h$.

Proof. The proof is immediate from Corollary 4.2.6 and (4.2.2).

This theorem shows one of the reasons for introducing cl_R.

4.3 Closed functions applied to normality and converse duality

If we are given a bifunction F and define

$$V = \{(y, \mu): \mu \geqslant F(x, y) \quad \text{for some } x\}$$

and $T = \{(y, \mu): y = 0\}$, then by definition F is normal if $\text{cl}\,(T \cap V) = T \cap (\text{cl}\, V)$. Thus normality is defined in terms of the closure of certain *sets*. In the next theorem this is translated to the closure of the perturbation function $p(y) = \inf_X F(x, y)$.

4.3.1 **Theorem.** Let a bifunction F be given and let V, T and p be as above. Then $\mathrm{cl}(T \cap V) = T \cap (\mathrm{cl}\, V)$ if and only if $(\mathrm{cl}\, p)(0) = p(0)$.

Proof. $(\mathrm{cl}\, p)(0) = \inf\{\mu : (0, \mu) \in \mathrm{cl\, epi}\, p\}$. By Lemma 3.5.6 this is equal to $\inf\{\mu : (0, \mu) \in T \cap \mathrm{cl}\, V\}$. On the other hand

$$\begin{aligned} p(0) &= \inf\{\mu : \mu \geqslant F(x, 0) \text{ for some } x\} \\ &= \inf\{\mu : (0, \mu) \in T \cap V\} = \inf\{\mu : (0, \mu) \in \mathrm{cl}(T \cap V)\}. \end{aligned}$$

The theorem follows immediately by observing that both $T \cap \mathrm{cl}\, V$ and $\mathrm{cl}(T \cap V)$ are closed.

Theorem 3.6.1. can now be formulated as follows. If F is convex and $(\alpha, \beta) \neq (+\infty, -\infty)$ then $\alpha = \beta$ if and only if $(\mathrm{cl}\, p)(0) = p(0)$. Taking cl_R rather than cl leads to a somewhat different result. If F is convex so is p and then $p^{**} = \mathrm{cl}_R p$ as follows from Theorem 4.2.9. In particular $\beta = p^{**}(0) = (\mathrm{cl}_R p)(0)$.

4.3.2 **Theorem.** Let F, p, α and β be as before and let F be convex. Then $\alpha = \beta$ if and only if $(\mathrm{cl}_R p)(0) = p(0)$.

In other words, if we call F *R-normal* when $(\mathrm{cl}_R p)(0) = p(0)$, then $\alpha = \beta$ if and only if F is *R-normal*, at least when F is convex.

Let us now pursue the implications of the equality $F = \mathrm{cl}_R F$ in which case we shall say that F is *R-closed.* Then by Theorem 4.2.9. we have that $F^{**} = F$, at least if F is convex, and by the definition of F^{**} it follows from this that

$$F(x, y) = \sup_{X^*, Y^*} \inf_{X, Y} (x^*x + y^*y - x^*x' - y^*y' + F(x', y')),$$

or if we change the sign of y^* and take $y = 0$,

$$\begin{aligned} \alpha &= \inf_X F(x, 0) \\ &= \inf_{x \in X} \sup_{X^*, Y^*} \inf_{x' \in X, Y} (x^*x - x^*x' + y^*y' + F(x', y')). \end{aligned}$$

On the other hand we have that

$$(-p^{\mathrm{d}})^{**}(0) = \sup_X \inf_{X^*} (-x^*x + (-p^{\mathrm{d}})(x^*)).$$

And if we replace $p^d(x^*)$ by $\sup_{Y^*} F^d(y^*, x^*)$ and $F^d(y^*, x^*)$ by $\inf_{X,Y}(F(x,y) - x^*x + y^*y)$ it is easily found that $(-p^d)^{**}(0) = -\alpha$. But $-p^d$ is convex if F is convex, so $-\alpha = (\mathrm{cl}_R - p^d)(0)$. Obviously $-\beta = (-p^d)(0)$ and so we obtain the following result.

4.3.3 **Theorem.** If F is convex and R-closed then $\alpha = \beta$ if and only if $(\mathrm{cl}_R - p^d)(0) = (-p^d)(0)$, hence if and only if $-F^d$ is R-normal.

We can get rid of the minus signs in this theorem if we introduce the *concave closure* of a function h by replacing the right-hand side of

$$(\mathrm{cl}\, h)(z) = \inf\{\nu\colon \nu \in R, (z, \nu) \in \mathrm{cl}\,\mathrm{epi}\, h\}$$

by

$$\sup\{\nu\colon \nu \in R, (z, \nu) \in \mathrm{cl}\,(-\mathrm{epi}\, h)\}$$

and changing the definition of cl_R appropriately. Here $-\mathrm{epi}\, h$ is the *hypograph* of h.

This theorem is important because the regularity condition that $-F^d$ should be R-normal is expressed in terms of the dual problem rather than the primal. All the duality theorems derived so far have counterparts for F convex and R-closed. In particular this is important if we consider the '*converse*' of a duality theorem stating among other things that a dual solution y_0^* exists, for then the corresponding converse duality theorem says among other things that a primal solution x_o exists and this, of course, is our primary interest. The next section will show other interesting consequences of converse duality.

Finally we remark that by Lemma 4.2.4 $\mathrm{cl}\, h$ is nowhere $-\infty$ if this function is finite somewhere. Then $\mathrm{cl}_R h = \mathrm{cl}\, h$, so that the property of being R-closed can be established by showing that the function under consideration is finite somewhere and is closed.

4.4 The perturbation function and subdifferentiability

In the very beginning we introduced the perturbation function $p(y) = \inf_X F(x, y)$ and wondered whether a y_0^* existed such that

(4.4.1) $$p(y) + y_0^* y \geqslant p(0) \quad \text{for all} \quad y \in Y.$$

This inequality led us directly to the heart of duality theory and subsequently grew out to the results of the foregoing sections of chapters 3 and 4. In 2.4, however, we hinted at the relation of (4.4.1) to differentiability, a relation which we have ignored so far, but which is very important because it will lead to the conclusion that in order to compute $\alpha = p(0)$ it is sufficient to know $p(y)$ for small values of y only.

It is inviting to try to derive from (4.4.1) a conclusion about the derivative of p at $y = 0$ by letting y tend to zero. Unfortunately p is not always differentiable at $y = 0$. As an example let $X = R_2$, $Y = R_3$, $x = (\xi_1, \xi_2)$, $y = (\eta_1, \eta_2, \eta_3)$, $F(x, y) = \xi_1 + \xi_2$ if $\xi_1 + \eta_1 \geqslant 1$, $\xi_1 + \xi_2 + \eta_3 \geqslant 2$, and $F(x, y) = +\infty$ otherwise. The feasible region of this simple linear programming problem is bounded by three hyperplanes all passing through the optimal point $x_0 = (1, 1)$. In a case like this we speak of *degeneration*. If $X = R_n$ degeneration means that at least $n+1$ hyperplanes bounding the feasible region have a certain point in common. One might object that degeneration is not likely to occur. For many problems this is true, but there are problems where a large range of y values must be examined. Then degeneration is quite likely to occur for certain values of y. Returning to our example, we see that it is not a difficult exercise to find that

$$p(y) = 2 - \eta_1 - \eta_2 \quad \text{if} \quad \eta_1 + \eta_2 \leqslant \eta_3$$

and that

$$p(y) = 2 - \eta_3 \quad \text{if} \quad \eta_1 + \eta_2 \geqslant \eta_3,$$

showing that p is not differentiable at $y = 0$.

Rather than letting y tend to zero, we replace y by τy, with $\tau \in R$, fix y and let τ tend to zero. Even then problems can arise unless we restrict τ to positive values. Hence instead of approaching $y = 0$ in an arbitrary manner we approach this point against the direction of a given y. This explains why we need the following definition.

4.4.2 **Definition.** Let Z be a linear space, h a function $h: Z \to [-\infty, +\infty]$, and $h(z_0)$ finite for some $z_0 \in Z$. Then $h'(z_0, z)$ is defined by

$$h'(z_0, z) = \lim_{\tau \downarrow 0} \frac{h(z_0 + \tau z) - h(z_0)}{\tau} \tag{4.4.3}$$

if the limit, which may be $+\infty$ or $-\infty$, exists. If $z \neq 0$ we call $h'(z_0, z)$ the *one-sided directional derivative of h at z_0 in the direction of z*.

It is convenient to introduce a separate symbol for the difference quotient in (4.4.3):

$$\text{(4.4.4)} \qquad \mathrm{D}h(z, \tau) = \frac{h(z_0 + \tau z) - h(z_0)}{\tau}.$$

Here we have suppressed z_0 since it is a fixed element of Z and often will be zero. Hence we can write

$$\text{(4.4.5)} \qquad h'(z_0, z) = \lim_{\tau \downarrow 0} \mathrm{D}h(z, \tau).$$

Let us call a function $h: Z \to [-\infty, +\infty]$ *positively homogeneous* (of the first degree) if $h(\mu z) = \mu h(z)$ for all $\mu > 0$; then we see immediately that h' is positively homogeneous in its second argument. This fact and some others are summarized in the following lemma

4.4.6 **Lemma.** If h is a convex function $h: Z \to [-\infty, +\infty]$ and if $h(z_0)$ is finite for some $z_0 \in Z$, then $\mathrm{D}h(z, \tau)$ is a nondecreasing function of τ if $\tau > 0$, so that $h'(z_0, z)$ exists for all $z \in Z$. Further, $h'(z_0, z)$ is convex and positively homogeneous in z, and

$$\text{(4.4.7)} \qquad -h'(z_0, -z) \leqslant h'(z_0, z).$$

Proof. For simplicity we take $z_0 = 0$. Let λ and μ be numbers such that $0 < \lambda\mu < \mu$, which implies that $0 < \lambda < 1$. By the convexity of h we have that

$$h(\lambda\mu z) \leqslant \lambda h(\mu z) + (1 - \lambda) h(0),$$

even if $h(\lambda\mu z) = +\infty$, for then $h(\mu z)$ must be $+\infty$ as well, as follows from the convexity of dom h and the fact that $h(0)$ is finite. It follows that $\mathrm{D}h(z, \lambda\mu) \leqslant \mathrm{D}h(z, \mu)$, hence $\mathrm{D}h(z, \tau)$ is nondecreasing in τ if $\tau > 0$.

To show (4.4.7) consider the inequality

$$h(0) \leqslant \tfrac{1}{2}h(-\tau z) + \tfrac{1}{2}h(\tau z) \quad \text{for} \quad \tau > 0.$$

Here we may assume that $h(\tau z) < +\infty$ if τ is small enough, for if not then $h'(0, z) = +\infty$ and (4.4.7) holds trivially. And we may also assume that $h(-\tau z) < +\infty$ if τ is small enough. Then it follows that $\mathrm{D}h(z, -\tau) \leqslant \mathrm{D}h(z, \tau)$ and (4.4.7) follows by letting τ tend to zero.

Before relating (4.4.1) and $p'(0, y)$ to each other we introduce the idea of a subgradient.

4.4.8 **Definition.** Let Z be a locally convex topological vector space with dual Z^*, and $h: Z \to [-\infty, +\infty]$ such that $h(z_0)$ is finite for some $z_0 \in Z$. Then $z^* \in Z^*$ is a *subgradient of h at* z_0 if

(4.4.9) $$h(z) - h(z_0) \geqslant z^*(z - z_0) \quad \text{for all} \quad z \in Z.$$

The set of all subgradients of h at z_0 is called the *subdifferential of h at* z_0 and is indicated by $\partial h(z_0)$.

4.4.10 **Theorem.** Let Z be a locally convex topological vector space with dual Z^* and let $h: Z \to [-\infty, +\infty]$ be a convex function such that $h(z_0)$ is finite for some $z_0 \in Z$. Define the function k by $k(z) = h'(z_0, -z)$. Then $-z_0^* \in \partial h(z_0)$ if and only if

(4.4.11)
$$k^*(z_0) = \sup_Z (z_0^* z - k(z)) = \inf_Z (z_0^* z + h'(z_0, z)) = 0.$$

Moreover we have that

(4.4.12) $$(\mathrm{cl}_R k)(z) = \sup\{z^* z\colon -z^* \in \partial h(z_0)\}.$$

Proof. For simplicity we again take $z_0 = 0$. Let $-z_0^* \in \partial h(0)$, so that $h(z) - h(0) \geqslant -z_0^* z$ for all z, hence $\mathrm{D}h(z, \tau) \geqslant -z_0^* z$ if $\tau > 0$ and also $\mathrm{D}h(-z, \tau) \geqslant z_0^* z$. The latter implies that $k(z) \geqslant z_0^* z$, hence that $\sup_Z (z_0^* z - k(z)) \leqslant 0$. But $k(0) = 0$, so that this supremum is equal to zero.

Conversely, let $\sup_Z (z_0^* z - k(z)) = 0$. Then $k(z) \geqslant z_0^* z$ for all z, and since $k(z)$ is the limit of $\mathrm{D}h(-z, \tau)$ and $\mathrm{D}h(-z, \tau)$ is nondecreasing as a function of τ, we also have $\mathrm{D}h(-z, \tau) \geqslant z_0^* z$ for all z and it easily follows that $-z_0^* \in \partial h(0)$.

The second part of the theorem is shown by observing that

$$\sup_Z (z_0^* z - k(z)) = -\infty \quad \text{when} \quad -z_0^* \notin \partial h(0);$$

for then $h(z) - h(0) \leqslant -z_0^* z - \delta$ for some $\delta > 0$ and some $z \in Z$, and since $\mathrm{D}h(z, \tau)$ is nondecreasing in its second argument,

$k(-z) = h'(0, z) \leqslant -z_\circ^* z - \delta$. But $k(z)$ is positively homogeneous and so is $k(-z) + z_\circ^* z$; hence indeed

$$\sup_Z (z_\circ^* z - k(z)) = \sup_Z (-z_\circ^* z - k(-z)) = +\infty.$$

By Theorem 4.2.9 we have

$$(\mathrm{cl}_R k)(z) = k^{**}(z) = \sup_{Z^*} (z^* z - \sup_Z (z^* z' - k(z'))),$$

and we may now restrict z^* to $-z^* \in \partial h(0)$, in which case, by the first part of the theorem, $\sup_Z (z^* z' - k(z')) = 0$, and (4.4.12) follows.

Notice that if (4.4.12) holds $\mathrm{cl}_R k$ is the support function of the set $-\partial h(z_\circ)$, which is the conjugate of the indicator of this set (see Definition 3.14.5).

4.4.13 **Corollary.** Let the bifunction $F: X \times Y \to [-\infty, +\infty]$ be given, with Y a locally convex topological vector space with dual Y^*. If $p(y) = \inf_X F(x, y)$ is convex (e.g. if F is convex) and if $p(0)$ is finite, then the following statements are equivalent

(4.4.14)

(*a*) $\quad \inf = \sup$ and $\sup$ is attained at $y_\circ^*$,

(*b*) $\quad y_\circ^* y + p(y) \geqslant p(0) \quad$ for all $\quad y \in Y$,

(*c*) $\quad \inf_Y (y_\circ^* y + p(y)) = p(0)$,

(*d*) $\quad -y_\circ^* \in \partial p(0)$,

(*e*) $\quad \inf_Y (y_\circ^* y + p'(0, y)) = 0$.

Proof. Immediate from foregoing results.

The next corollary deals with $y_\circ^*$ being the unique dual solution.

4.4.15 **Corollary.** Let everything be as in the preceding corollary. Also assume that $q(y) = p'(0, -y)$ is R-closed. Then $y_\circ^*$ is the *unique* dual solution if and only if q is a continuous linear functional from Y to R, in which case $\lim_{\tau \to 0} \mathrm{D}p(y, \tau)$ exists and is finite.

Proof. If $y_\circ^*$ is the unique dual solution then

$$q(y) = (\mathrm{cl}_R q)(y) = y_\circ^* y$$

by (4.4.12). Conversely, if $q(y) = y_0^* y$ for some $y_0^* \in Y^*$ then

$$y_0^* y = \sup\{y^* y : -y^* \in \partial p(0)\}$$

and necessarily $\partial p(0) = \{-y_0^*\}$, so that y_0^* is unique.

When taking in this corollary the limit of $\mathrm{D}p(y, \tau)$, τ is not restricted to positive values, hence we obtain a *two*-sided directional derivative of p at $y = 0$. This means that if q is a continuous linear functional then p is *Gâteaux differentiable* at $y = 0$.

As we only used the fact that k was R-closed in order to conclude that $\mathrm{cl}_R k = k^{**}$, we may replace $\mathrm{cl}_R k$ by $\mathrm{cl}\, k$ if we can show that $\mathrm{cl}\, k = k^{**}$. Theorem 4.2.8 provides sufficient conditions for this, but it may not be so easy to verify these conditions in a practical situation. A similar remark applies to verifying that k is R-closed.

If in the corollaries we assume that $\partial p(0)$ is not empty, then $q(y) \geqslant y_0^* y$ for some y_0^*, hence $(\mathrm{cl}\, q)(y) \geqslant y_0^* y > -\infty$ so that $\mathrm{cl}_R q = \mathrm{cl}\, q$.

4.4.16 **Example.** Let $Y = R$ and suppose that $p(y) = 1 + 2y$ if $y \geqslant 0$ and $p(y) = 1 - 3y$ if $y \leqslant 0$. Then $\partial p(0)$ is the closed interval $[-3, 2]$, and $p'(0, y) = 2y$ if $y \geqslant 0$ and $p'(0, y) = -3y$ if $y \leqslant 0$. Each y_0^* such that $-y_0^* \in \partial p(0)$ leads to a hyperplane $\{(y, \mu): y_0^* y + \mu = p(0)\}$ supporting $\mathrm{epi}\, p$ at $(0, p(0)) = (0, 1)$. We leave it to the reader to draw a picture illustrating these facts.

If X as well as Y is a locally convex topological vector space and F is convex and R-closed as a function of $(x, y) \in X \times Y$, then we are allowed to translate the foregoing duality results into results about converse duality.

4.4.17 **Corollary.** Let everything be as in Corollary 4.4.13. In addition assume that X is a locally convex topological vector space with dual X^*, and that F is convex, so that $-p^{\mathrm{d}}$ is also convex. If F is R-closed then the following statements are equivalent

(4.4.18)

(*a*) inf = sup and inf is attained at x_0,

(*b*) $x^*x_o + p^d(x^*) \leq p^d(0)$ for all $x^* \in X^*$,

(*c*) $\sup_{X^*}(x^*x_o + p^d(x^*)) = p^d(0)$,

(*d*) $x_o \in \partial(-p^d)(0)$,

(*e*) $\sup_{X^*}(x^*x_o + (p^d)'(0, x^*)) = 0$.

4.4.19 **Corollary.** Let the assumptions of Corollary 4.4.17 be true and let $(-p^d)'(0, x^*)$ be R-closed. Then x_o is the ***unique*** primal solution if and only if this function is a continuous linear functional from X^* to R.

We see from the results of this section that in general $p(y)$ need only be known for small values of y if we want to compute all optimal solutions y_0^* of the dual problem, and that $p^d(x^*)$ need only be known for small values of x^* if we want to compute all optimal solutions x_o of the primal problem. The word 'small' should be carefully understood here, because in one direction it might be necessary to consider much 'smaller' values than in another direction. This is the consequence of the relevant functions being only directionally differentiable or even one-sidedly directional.

4.5 The Lagrangian

The discussion of this section is a continuation of that in 3.10, where we saw that the part played by the Lagrangian involving only the primal problem is symmetric to the one it plays involving only the dual problem at least if

(4.5.1) $$F(x, y) = \sup_{Y^*}(L(x, y^*) - y^*y).$$

But since by definition

(4.5.2) $$L(x, y^*) = \inf_Y (F(x, y) + y^*y)$$

this condition is equivalent to

(4.5.3) $$F(x, y) = \sup_{Y^*} \inf_Y (F(x, y') + y^*y' - y^*y)$$

which, as noticed already in 4.1, says that the biconjugate of the function $F(x, \cdot)$ is that function itself for all $x \in X$. Combining this with Theorem 4.2.9 we have the following result.

4.5.4 **Theorem.** Let a bifunction $F: X \times Y \to [-\infty, +\infty]$ be given. If the function $F(x, \cdot): Y \to [-\infty, +\infty]$ is convex and R-closed for each $x \in X$, then $F(x, y) = \sup_{Y^*} (L(x, y) - y^*y)$.

As an example let F be as in 3.11, i.e. let

$$(4.5.5) \qquad \begin{aligned} F(x,y) &= f(x) \quad \text{if} \quad g(x) \leqslant y \quad \text{and} \quad x \in C, \\ &= +\infty \quad \text{otherwise,} \end{aligned}$$

and assume that f and g are convex functions from the convex set C to R and Y, respectively. Also assume that the positive cone of P is a closed set. Then obviously $F(x, \cdot)$ is a closed function which is nowhere $-\infty$ so that it is R-closed as well. Hence F can be computed from (4.5.1), where now

$$(4.5.6) \quad \begin{aligned} L(x,y^*) &= f(x) + y^*g(x) && \text{if} \quad x \in C \quad \text{and} \quad y^* \geqslant 0, \\ &= -\infty && \text{if} \quad x \in C \quad \text{and} \quad y^* \ngeqslant 0, \\ &= +\infty && \text{if} \quad x \notin C. \end{aligned}$$

When the constraint $g(x) \leqslant y$ is replaced by $Bx = b + y$, as in 3.12, then again we may apply (4.5.1).

Next consider the case of Fenchel duality, treated in 3.14, so that

$$(4.5.7) \qquad F(x,y) = f(x) + g(x+y)$$

with f and g convex functions from X to $(-\infty, +\infty]$. The fact that $F(x, \cdot)$ is R-closed now simply follows from the fact that q is. In particular if g is the indicator function of a convex set G, $g(y) = \delta(y{:}\ G)$, we see that $F(x, \cdot)$ is R-closed if and only if G is a closed set.

These examples show that we may be confident that in many practical cases F can be recovered from L, because usually one assumes that positive cones and other relevant sets are closed, and it it is not possible to assume that a certain set is closed one can sometimes deduce this from the assumptions made. For these reasons it is important to consider Theorems 3.10.8 and 3.10.9 because the existence of a saddle-point or a saddle-value of L may immediately lead to duality results. These results may hold even if F is not a convex function, although, as we have seen in 3.10, $F(x, \cdot)$ must be convex for each $x \in X$.

Another reason to study the existence of saddle-values and saddle-points of a given function $L(x, y^*)$ is that we might not be interested so much in finding $\inf_X f(x, 0)$, that is in solving a certain optimization problem, as in solving a so-called *game* problem. An extremely simple game problem is given in Example 1.2.7, where $x_1 = x$, $x_2 = y^*$ and $\phi(x_1, x_2) = L(x, y^*)$, so that L becomes the 'pay-off function' of the game.

4.6 Saddle-points, variational inequalities and the complementarity problem

Given a bifunction $F: X \times Y \to [-\infty, +\infty]$, the equality of the infimum of the corresponding primal problem and the supremum of the corresponding dual problem together with the solvability of both the primal and the dual problem leads, as we have seen, to the existence of a saddle-point (x_o, y_0^*) of $L(x, y^*) = \inf_Y (F(x, y) + y^*y)$, i.e.

(4.6.1)

$$\begin{aligned}\inf_Y (F(x_o, y) + y^*y) &\leqslant \inf_Y (F(x_o, y) + y_0^*y) \\ &\leqslant \inf_Y (F(x, y) + y_0^*y) \quad \text{for all} \quad (x, y^*).\end{aligned}$$

Nothing is lost if we leave out the expression in the middle:

(4.6.2)

$$\inf_Y (F(x_o, y) + y^*y) \leqslant \inf_Y (F(x, y) + y_0^*y) \quad \text{for all} \quad (x, y^*).$$

Specializing to *extended Fenchel duality*, where

$$F(x, y) = f(x) + g(Ax + y)$$

we obtain from (4.6.2)

(4.6.3)

$$y_0^*Ax - y^*Ax_o \leqslant f(x) + g^*(-y^*) - [f(x_o) + g^*(-y_0^*)] \quad \text{for all} \quad (x, y^*).$$

If we define $A^*: Y^* \to X^*$ by $(A^*y^*) = y^*(Ax)$, and if we let $z = (x, y^*)$, $z_0 = (x_o, y_0^*)$, $Bz = (-A^*y^*, Ax)$, $\langle (x^*, y), (x, y^*) \rangle = x^*x + y^*y$, so that $\langle Bz_o, z_0 \rangle = 0$, and let $h(z) = f(x) + g^*(-y^*)$, then (4.6.3) takes the form

$$\langle Bz_o, z_0 - z \rangle \leqslant h(z) - h(z_0) \quad \text{for all} \quad z. \qquad (4.6.4)$$

This inequality, which is only meaningful if $h(z_0)$ is *finite*, is also known as a *variational inequality*. Since we have equality if $z = z_0$, we can also write

$$\inf_Z (h(z) + \langle Bz_o, z - z_0 \rangle) = h(z_o). \tag{4.6.5}$$

Remark. This equation shows some similarity with $\inf_Y (p(y) + y_0^* y) = p(0)$, our starting point for duality theory. But beware because here $y_0 = 0$ is known, whereas in (4.6.5) $z_0 = (x_o, y_0^*)$ is not known and moreover Bz_o, playing the part of y_0^*, depends on z_0.

Obviously, finding a saddle-point of the Lagrangian for the extended Fenchel duality problem amounts to solving the variational inequality (4.6.3). This can be written as two inequalities by taking either $x = x_o$ or $y^* = y_0^*$, namely

$$y_0^* Ax_o - y^* Ax_o \leqslant g^*(-y^*) - g^*(-y_0^*) \quad \text{for all} \quad y^* \tag{4.6.6}$$

and

$$y_0^* Ax - y_0^* Ax_o \leqslant f(x) - f(x_o) \quad \text{for all} \quad x, \tag{4.6.7}$$

and the saddle-point condition for $L(x, y^*) = f(x) - y^* Ax - g^*(-y^*)$ immediately follows from this. Equation (4.6.7) is, of course, nothing but (3.14.16) for the case where $Y = X$ and A is the identity, and then (4.6.6) is similar to (3.14.17). Hence we have not really done anything new.

An interesting special case is obtained if we take $X = R_n$, $Y = R_m$, Q a symmetric $n \times n$ matrix, A an $n \times m$ matrix, $q \in R_n$, $b \in R_m$, $P = \{(y: y \geqslant 0\}$ and $S = \{x: x \geqslant 0\}$; and define f and g by

$$f(x) = q^t x + x^t Q x, \quad g(y, z) = \delta(y - b: P) + \delta(z: S),$$

where the superscript t denotes transposition (x and q are considered to be column vectors, as are y, z and b). Finally let

$$F(x, (y, z)) = f(x) + g(Ax + y, x + z)$$

which means that we have replaced y by (y, z) and A by (A, I) with I the identity matrix. Then the corresponding optimization problem is that of finding

$$\inf_X F(x, (0, 0)) = \inf \{q^t x + x^t Q x: Ax \geqslant b, x \geqslant 0\}$$

which is *quadratic programming*. Rather than considering this as a Lagrangian duality problem (which we leave to the reader which might yield the final result more quickly) we stick to our point of view of considering it as an extended Fenchel duality problem.

We remark that there is an alternative way to arrive at quadratic programming, i.e. by taking $f(x) = q^t x + x^t Qx + \delta(x: S)$ and $g(y) = \delta(y-b: P)$. Then, however, we obtain only one multiplier, namely for the constraint $Ax \geqslant b$, but none for the constraint $x \geqslant 0$ (unless some additional trickery is used), whereas we shall automatically find two multipliers if we proceed as indicated. This incidentally gives us a simple rule of thumb: if an inequality is included in g (by means of a suitable indicator function) then a multiplier for this constraint is obtained automatically, but not so if it is included in f.

Instead of (4.6.6) and (4.6.7) we get, using the definition of a conjugate function,

(4.6.8)

$$y_o^* Ax_o + z_o^* x_o - y^* Ax_o - z^* x_o \leqslant -\inf\{y^* y + z^* z: y \geqslant b, z \geqslant 0\} + \inf\{y_o^* y + z_o^* z: y \geqslant b, z \geqslant 0\} \quad \text{for all} \quad (y^*, z^*)$$

and

(4.6.9)

$$y_o^* Ax + z_o^* x - y_o^* Ax_o - z_o^* x_o \leqslant q^t x + x^t Qx - q^t x_o - x_o^t Qx_o \text{ for all } x.$$

Since the second infimum in (4.6.8) cannot be $-\infty$, we have that $y_o^* \geqslant 0$ and $z_o^* \geqslant 0$ and it follows that

$$y_o^* Ax_o + z_o^* x_o - y^* Ax_o - z^* x_o \leqslant -y^* b + y_o^* b \quad \text{for all} \quad y^* \geqslant 0, z^* \geqslant 0.$$

From this it is easily seen that $s_o = Ax_o - b \geqslant 0$ and $x_o \geqslant 0$, and moreover that $y_o^*(Ax_o - b) = 0$ and $z_o^* x_o = 0$ (which is complementary slackness). And from (4.6.9) we get that

$$y_o^* A + z_o^* = q^t + 2x_o^t Q \quad \text{or} \quad A^t y_o^{*t} + z_o^{*t} = q = 2Qx_o.$$

All this can now be neatly be summarized as follows,

$$(4.6.10)\quad \begin{pmatrix} z_0^{*t} \\ s_0 \end{pmatrix} = \begin{pmatrix} q \\ -b \end{pmatrix} + \begin{pmatrix} 2Q & -A^t \\ A & 0 \end{pmatrix} \begin{pmatrix} x_o \\ y_0^{*t} \end{pmatrix},$$

$$\begin{pmatrix} z_0^{*t} \\ s_0 \end{pmatrix} \geqslant 0, \quad \begin{pmatrix} x_o \\ y_0^{*t} \end{pmatrix} \geqslant 0, \quad (x_o^t, y_0^*) \begin{pmatrix} z_0^{*t} \\ s_0 \end{pmatrix} = 0.$$

This clearly has the following form, where the meaning of u_o, v_o, p and M is obvious,

$$(4.6.11)\qquad u_o = p + Mv_o, \quad u_o \geqslant 0, \quad v_o \geqslant 0, \quad u_o^t v_o = 0.$$

The problem of finding u_o and v_o such that (4.6.11) holds is called the (*linear*) *complementarity problem*. It so happens that this problem is equivalent to the problem of solving a certain variational inequality! To show this let $Bv = p + Mv$, then (4.6.11) becomes

$$(4.6.12)\qquad u_o = Bv_o, \quad u_o \geqslant 0, \quad v_o \geqslant 0, \quad u_o^t v_o = 0$$

and this implies

$$(4.6.13)\qquad v_o \geqslant 0, \quad \langle Bv_o, v_o \rangle \leqslant 0 \quad \text{for all} \quad v \geqslant 0,$$

for if $u_o = Bv_o$ and $u_o^t v_o = 0$ then $\langle Bv_o, v_o \rangle = 0$, and $\langle Bv_o, v \rangle \geqslant 0$ for all $v \geqslant 0$ if $u_o \geqslant 0$. Conversely, if (4.6.13) is true, then with $v = 0$ we obtain $\langle Bv_o, v_o \rangle \leqslant 0$, and with $v = 2v_o$, $\langle Bv_o, v_o \rangle \geqslant 0$, so that $\langle Bv_o, v_o \rangle = 0$ and $\langle Bv_o, v \rangle \geqslant 0$ for all $v \geqslant 0$, hence $Bv_o \geqslant 0$. Taking $u_o = Bv_o$ we find (4.6.12).

To see that (4.6.13) is a variational inequality, in (4.6.4) let $z = v$ and $h(v) = \delta(v\colon C)$ with $C = \{v\colon v \geqslant 0\}$.

The conclusion then is that variational inequalities can be obtained from the saddle-point condition of Lagrangians, and that the complementarity problem is equivalent to solving a special type of variational inequality. Both the complementarity problem and the variational inequality can be generalized by taking B more general. Quite a few research papers have been devoted to either establishing the existence of solutions or devising algorithms for finding solutions.

We remark that the variational inequality (4.6.4) is related to the defining equation of a subgradient z_0^* of h at z_0, namely

$$(4.6.14)\qquad z_0^*(z - z_0) \leqslant h(z) - h(z_0) \quad \text{for all} \quad z.$$

For solving the variational inequality means finding z_0 and a subgradient z_0^* of h at z_0 such that z_0^* has the special form $-Bz_o$.

Although the relationships found are interesting, our curiosity is not yet satisfied, because we have only a complementarity problem relating to a very special problem, i.e. quadratic programming: if x^tQx is convex, if the infimum of this problem is finite and if an appropriate constraint qualification is satisfied, then the complementarity problem has a solution if and only if inf = min = sup = max. Does there exist a system of (in)equalities for an arbitrary extended Fenchel duality problem, say to find $\inf_X (f(x)+g(Ax))$, such that this system has a solution (x_o, y_o^*) if and only if

$$\inf = f(x_o)+g(Ax_o) = \inf_X (f(x)-y_o^*Ax) + \inf_Y (g(y)+y_o^*y) = \sup?$$

The answer is in the affirmative and we have almost written down the required system, for the latter equality means that

$$f(x_o)+g(Ax_o) \leqslant f(x)-y_o^*Ax+g(y)+y_o^*u \quad \text{for all} \quad (x, y),$$

and taking $y = Ax_o$ or $x = x_o$ we get the desired result,

(4.6.15)
$$f(x_o)+f^*(y_o^*A)-y_o^*Ax_o = 0$$
and
$$g(Ax_o)+g^*(-y_o^*)+y_o^*Ax_o = 0.$$

Nothing prevents us from considering this as a *generalization of complementarity*. Indeed, if we let $f(x) = q^tx+x^tQx$ and $g(y, x) = \delta(y-b: P)+\delta(z: Q)$, again replacing y by (y, z) and A by (A, I), then (4.6.10) can be found from (4.6.15).

Since, as we have seen, Fenchel duality is just as general as bifunctional duality, we can restate the derived results in terms of a given bifunction F. Then we must introduce the sets $V = \{(y, \mu): \mu \geqslant F(x, y) \text{ for some } x\}$ and $T = \{(y, \mu): y = 0\}$, and define f and g by $f(y, \mu) = \mu$ if $(y, \mu) \in \operatorname{dom} f = T$ and $g(y, \mu) = 0$ if $(y, \mu) \in \operatorname{dom} g = V$ (see 3.16). We shall not, however, work this out but leave the details to the reader.

5

A local approach for optimization problems in Banach spaces

5.1 Introduction; Fréchet differentiability; some lemmas

In 3.13 we learnt that if $\alpha = \inf\{f(x): g(x) \leqslant 0,\ h(x) = 0,\ x \in C\}$ is finite, if $f\colon X \to R$ and $g\colon X \to Y$ are convex functions and $h\colon X \to Z$ is an affine function, and if C is a convex set, then under certain regularity conditions (see (3.13.4))

$$\alpha = \inf\{f(x)+y_0^*g(x)+z_0^*h(x): x \in C\} \quad \text{for some } y_0^* \geqslant 0 \text{ and some } z_0^*, \tag{5.1.1}$$

and that if x_0 is a primal solution then $y_0^*g(x_0) = 0$.

Now suppose that $x_0 \in \operatorname{int} C$, or that C is open, or that $C = X$. At any rate assume that C really is superfluous. Then if X, Y and Z are finite-dimensional and f, g and h are differentiable we would have that

$$\nabla_x(f(x_0)+y_0^*g(x_0)+z_0^*h(x_0)) = 0 \quad \text{for some } y_0^* \geqslant 0 \text{ and some } z_0^*, \tag{5.1.2}$$

and, of course, that

$$y_0^*g(x_0) = 0, \quad g(x_0) \leqslant 0, \quad h(x_0) = 0. \tag{5.1.3}$$

Conditions (5.1.2) and (5.1.3) together are known as the *Kuhn–Tucker conditions* or the necessary Kuhn–Tucker optimality conditions.

What changes are necessary if we let C play a more interesting part? Clearly we run the risk that x_0 will be on the boundary of C and then (5.1.2) in general will not hold. Fenchel duality now comes to the rescue. Still assuming that (5.1.1) is true, let

$$\tilde{f}(x) = f(x)+y_0^*g(x)+z_0^*h(x);$$

then we get $\alpha = \bar{f}(x_o) = \inf\{\bar{f}(x): x \in C\}$ and if the Fenchel duality theorem (3.14.12) is applicable, then by (3.14.16) and (3.14.18) $f(x) - f(x_o) + y_o^*(g(x) - g(x_o)) + z_o^*(h(x) - h(x_o)) \geq x_o^*(x - x_o)$ for some x_o^* and all $x \in X$, and $x_o^*(x - x_o) \geq 0$ for all $x \in C$. From the first inequality it would follow that

$$\nabla_x (f(x_o) + y_o^* g(x_o) + z_o^* h(x_o))(x - x_o) = x_o^*(x - x_o) \text{ for all } x \in X,$$

and combining this with the second one we would have that

(5.1.4)
$$\nabla_x (f(x_o) + y_o^* g(x_o) + z_o^* h(x_o))(x - x_o) \geq 0 \quad \text{for all} \quad x \in C.$$

We shall see in 7.3 that a special case of (5.1.4) leads to the so-called *minimum principle* of optimal control, and this is why several authors call (5.1.4) itself the minimum principle or the pre-minimum principle of the given problem.

Clearly if x_o is not on the boundary of C then it is one simple step from (5.1.4) to (5.1.2).

Whether or not (5.1.2) is true is determined solely by the behaviour of f, g and h in the vicinity of x_o, i.e. by *local* properties of these functions, local with respect to x_o, that is. So the natural question arises whether the convexity assumptions about f and g and the assumption that h must be affine, assumptions which are *global* in nature, are really important for (5.1.2) to hold. A similar question can be posed as far as (5.1.4) is concerned, but we should observe that here we have a mixture of local and global aspects in that x may vary over the entire set C. The answer to these questions is that in general the assumptions are not necessary, but that we cannot forget about the convexity of C (unless C is '*linearized*' in some way or other as will be done in 5.4 and 5.5). This is extremely important because there exist quite a few nonconvex optimization problems (but with convex C) involving differentiable functions. So let us now focus our attention on differentiability. But what type of differentiability? Although interesting results can be obtained from the more general Gâteaux differentiability (see 4.4), we shall restrict ourselves to Fréchet differentiability, which requires that the relevant spaces must be Banach spaces or at least normed spaces.

5.1.5 **Definition.** We say that a function q from a Banach space X to another Banach space $\tilde{X}$ is *Fréchet differentiable at* $x_o \in X$ if there exists a continuous linear function $q'(x_o)\colon X \to \tilde{X}$ such that

$$\lim_{x \to 0} \frac{|q(x_o+x)-q(x_o)-q'(x_o)x|}{|x|} = 0.$$

The function q is said to be *Fréchet differentiable in* X if q is Fréchet differentiable at x for all $x \in X$; $q'(x_o)$ is called the *Fréchet derivative of* q *at* x_o and $q'(x_o)x$ is called the *Fréchet differential at* x_o *with increment* x.

When $X = \tilde{X} = R$ we simply have that $q'(x_o)x = (\mathrm{d}q(x_o)/\mathrm{d}x)x$. It should be noticed that although in this case everybody will interpret $q'(x_o)$ as a *number* by which x is to be multiplied, in general $q'(x_o)$ is a *mapping* (from X to $\tilde{X}$). The consequence of this is that $q'\colon x_o \mapsto q'(x_o)$ is a mapping from X to the normed space $L(X, \tilde{X})$ of all continuous linear mappings from X to $\tilde{X}$, and that the *second* Fréchet derivative $q''(x_o)$ of q at x_o is a continuous linear mapping from X to $L(X, \tilde{X})$ and therefore is an element of $L(X, L(X, \tilde{X}))$, etc.

The proof of the following lemma is similar to the proof of a corresponding lemma in elementary analysis and is therefore left to the reader.

5.1.6 **Lemma**

(*a*) If $q'(x_o)$ exists it is unique and q is continuous at x_o.

(*b*) If X possesses a positive cone P with nonempty interior, and $x_o, x \in X$ are such that $q'(x_o)x \in \operatorname{int}(-P)$, then $q(x_o+\tau x)-q(x_o) \in \operatorname{int}(-P)$ if $\tau > 0$ and small enough.

(*c*) If P is as under (*b*) and x_o, $x \in X$ are such that $q(x_o) \in -P$ and $q(x_o)+q'(x_o)x \in \operatorname{int}(-P)$ then $q(x_o+\tau x \in \operatorname{int}(-P)$ if $\tau > 0$ and small enough.

(*d*) Given x_o and x, we have

$$q(x_o+x)-q(x_o) = q'(x_o+\theta_1 x)x$$

and $$q(x_o+x)-q(x_o)-q'(x_o)x = \tfrac{1}{2}q''(x_o+\theta_2 x)xx$$

for some $\theta_{1,2}$, $0 < \theta_{1,2} < 1$ (assuming that q' and q'' exist, and denoting $(q''(x_o)x)x$ by $q''(x_o)xx$).

Apart from these elementary facts we shall need what might be called a *linearization lemma.*

5.1.7 **Lemma.** Let q be a mapping from a Banach space X to another Banach space $\tilde{X}$, let q be Fréchet differentiable in a neighbourhood of a point $x_o \in X$ and let q' be continuous at x_o. Further, assume that $q(x_o) = 0$, that $q'(x_o)$ is a mapping *onto* $\tilde{X}$ and that $q'(x_o)x_1 = 0$ for some $x_1 \in X$. Then

$$q(x_o + \lambda x_1 + o(\lambda)) = 0, \tag{5.1.8}$$

where $o(\lambda)$ is an order-of function, i.e. a function such that

$$\lim_{\lambda \to 0} \frac{|o(\lambda)|}{\lambda} = 0. \tag{5.1.9}$$

Proof. See Appendix E.

Equation (5.1.8) says that for a suitable neighbourhood U of the origin of X the set $\{x: q(x_o + x) = 0,\ x \in U\}$ can be approximated by the intersection of U and the line $\{\lambda x_1: \lambda \in R\}$, in other words, that a small portion of the curve $\{x: q(x_o + x) = 0\}$ can be approximated by a line segment, which explains the term 'linearization lemma'. We shall use the lemma when we come to an equality constraint of the form $q(x) = 0$. Then from (5.1.8) we may infer that $x_o + \lambda x_1 + o(\lambda)$ is feasible and draw conclusions from that.

With differentiability the definition of convexity can be given another form.

5.1.10 **Lemma.** Let q be a Fréchet differentiable function from a Banach space X to another Banach space $\tilde{X}$ with positive cone. Then q is convex if and only if

$$q'(x_1)(x_2 - x_1) \leqslant q(x_2) - q(x_1) \text{ for all } x_1, x_2 \in X. \tag{5.1.11}$$

Proof. Let q be convex and take λ such that $0 < \lambda < 1$. If $x_1 = x_2$ there is nothing to prove, so let $x_1 \neq x_2$. By convexity

$$\frac{q(x_1 + \lambda(x_2 - x_1)) - q(x_1)}{\lambda} \leqslant q(x_2) - q(x_1).$$

If we let λ tend to zero it follows from the definition of q' and the fact that $x_1 \neq x_2$ that (5.1.11) is true.

Conversely let $x_1, x_2 \in X$ be given and let λ be such that $0 < \lambda < 1$. Then by (5.1.11)

$$q'(\lambda x_1+(1-\lambda)x_2)(1-\lambda)(x_1-x_2) \leqslant q(x_1)-q(\lambda x_1+(1-\lambda)x_2)$$

and

$$q'(\lambda x_1+(1-\lambda)x_2)\,\lambda(x_2-x_1) \leqslant q(x_2)-q(\lambda x_1+(1-\lambda)x_2).$$

Multiplying these inequalities by λ and $1-\lambda$, respectively, and adding the results gives

$$0 \leqslant \lambda q(x_1)+(1-\lambda)q(x_2)-q(\lambda x_1+(1-\lambda)x_2).$$

If q is only differentiable at x_o then from convexity it follows that

(5.1.12) $$q'(x_o)(x-x_o) \leqslant q(x)-q(x_o) \text{ for all } x \in X.$$

5.2 Finitely many constraints; first-order conditions

Suppose we are given the problem of finding $\alpha = \inf\{f(x): g(x) \leqslant 0\}$ with f and g differentiable. When $x \in X = R_n$ and $g(x) \in Y = R_m$ we would like to conclude that

(5.2.1)
$$\nabla(f(x_o)+y_o^* g(x_o)) = 0, \quad y_0^* g(x_o) = 0 \quad \text{for some} \quad y_0^* \geqslant 0,$$

where ∇ is differentiation with respect to x. Let $g_i(x) \leqslant 0$ be the ith scalar constraint of $g(x) \leqslant 0$, $i = 1, \ldots, m$. Clearly if g_i is inactive at x_o, that is if $b_i(x_o) < 0$, then the ith component of y_0^* must be zero if $y_0^* g(x_o) = 0$ is to hold. Including all active constraints in g_a, say, it follows that for all x

(5.2.2) $$\nabla g_a(x_o)x \leqslant 0 \quad \text{implies that} \quad \nabla f(x_o)x \geqslant 0$$

if (5.2.1) is true. This can be seen by simply postmultiplying (5.2.1) by x. But what is also important, if conversely (5.2.2) is true for all x then (5.2.1) must hold. If we set equal to zero the components of y_0^* which correspond to the inactive constraints then this follows immediately from the next lemma if we let $c = \nabla f(x_o)$ and $A = \nabla g_a(x_o)$.

5.2.3 **Lemma** (Farkas 1902). If $c \in R_n$, A is an $m \times n$ matrix and for all $x \in R_n$ $Ax \leqslant 0$ implies that $cx \geqslant 0$, then $c + y_a^* A = 0$ for some $y_a^* \geqslant 0$. Here cx, Ax and $y_a^* A$ are matrix products.

Proof. Consider the linear programming problem of finding $\inf\{cx: Ax \leqslant 0\}$. Obviously this infimum is equal to zero and by Theorem 3.13.8 it follows that $c + y_a^* A = 0$ for some $y_a^* \geqslant 0$.

We have thus found a *necessary and sufficient regularity condition* for (5.2.1) to hold.

5.2.4 **Theorem.** Let $f\colon R_n \to R$ and $g\colon R_n \to R_m$ be differentiable functions and let x_o be a solution of the problem of finding $\inf\{f(x): g(x) \leqslant 0\}$. Then (5.2.1) holds if and only if the implication of (5.2.2) is true for all x.

This theorem can easily be applied to *equality constraints* as well; for if $h(x) = 0$ is such a constraint all we have to do is replace it by $h(x) \leqslant 0$ and $-h(x) \leqslant 0$. Since $h_a = h$, (5.2.2) then becomes

(5.2.5) $$\nabla h(x_o)x = 0 \quad \text{implies that} \quad \nabla f(x_o)x = 0.$$

Unfortunately our regularity condition (involving g as well as f) is not always satisfied.

5.2.6 **Counterexample.** Let $X = Y = R_2$, $x = (\xi_1, \xi_2)$, $f(x) = -\xi_1 - \xi_2$, $g_1(x) = \xi_2 + \xi_1^3$ and $g_2(x) = -\xi_2 + \xi_1^3$. Then $x_o = 0$ and both constraints are active at x_o. It is easily verified that $\nabla g(x_o)x \leqslant 0$ is equivalent to $\xi_2 = 0$ that $\nabla f(x_o)x = -\xi_1 - \xi_2 < 0$ if $\xi_1 > 0$ and $\xi_2 = 0$.

One might argue that such examples are rather abnormal. This is true but there is a more serious objection against (5.2.2); we ought to know x_o, for certainly (5.2.2) will in general not hold for *any* x_o. A sufficient condition for (5.2.2) to hold which does not involve x_o itself and moreover does not involve f, is Slater's constraint qualification

(5.2.7) $$g \text{ is convex and } \quad g(\hat{x}) < 0 \quad \text{for some } \hat{x} \in X.$$

5.2.8 **Theorem.** Let f and g be as before and let x_o be a solution of the problem of finding $\inf\{f(x): g(x) \leqslant 0\}$. Then (5.2.2), and hence the Kuhn–Tucker conditions (5.2.1), hold if (5.2.7) is true.

Proof. Suppose to the contrary that for some x $\nabla g_a(x_o)x \leqslant 0$ whereas $\nabla f(x_o)x < 0$. Fix $\sigma > 0$ such that $\nabla f(x_o)(x+\sigma(\hat{x}-x_o)) < 0$. From the convexity of g and (5.1.12) it follows that

$$\nabla g_a(x_o)(\hat{x}-x_o) \leqslant g_a(\hat{x})-g_a(x_o) = g_a(\hat{x}) < 0,$$

hence $\nabla g_a(x_o)(x+\sigma(\hat{x}-x_o)) < 0$. Setting $\tilde{x} = x+\sigma(\hat{x}-x_o)$ we have therefore that $\nabla g_a(x_o)\tilde{x} < 0$ and that $\nabla f(x_o)\tilde{x} < 0$. By Lemma 5.1.6(*b*) and (*c*), it follows from this that $f(x_o+\tau\tilde{x}) < f(x_o)$ and $g_a(x_o+\tau\tilde{x}) < 0$ if $\tau > 0$ and small enough. Since by Lemma 5.6.1(*a*) g is continuous at x_o we can take τ so small that $g_i(x_o+\tau\tilde{x}) < 0$ if g_i is inactive at x_o. Then $f(x_o+\tau\tilde{x}) < f(x_o)$ whereas $g(x_o+\tau\tilde{x}) < 0$, so that $x_o+\tau\tilde{x}$ is feasible *and* is better than optimal, which is impossible.

Notice that we cannot apply Slater's constraint qualification for equality constraints. The following approach is equally applicable to inequality constraints as well as equality constraints and involves the so-called cone of tangents of the set $G_a = \{x\colon g_a(x) \leqslant 0\}$, or if as well as the constraint $g(x) \leqslant 0$ a constraint $h(x) = 0$ is also involved, $G_a = \{x\colon g_a(x) \leqslant 0,\ h(x) = 0\}$. Hence G_a is the feasible region if inactive constraints are ignored.

5.2.9 **Definition.** The *cone of tangents of* G_a *at* x_o is defined by

(5.2.10) $$T = \{x\colon x = \lim_{n\to\infty} \lambda_n(x_n-x_o) \text{ such that } \lambda_n > 0,\quad x_n \in G_a,\quad x_n \to x_o\}.$$

T is, of course, a cone, but not necessarily a convex cone. Further let S be defined by

(5.2.11) $$S = \{x\colon \nabla g_a(x_o)x \leqslant 0,\quad \nabla h(x_o)x = 0\}.$$

In what follows we shall suppress h, because the present argument can be applied to equality constraints by representing them as two inequality constraints.

5.2.12 **Lemma.** $T \subset S$.

Proof. Trivially $0 \in T \cap S$, so let $x \in T$ and $x \neq 0$. Let x_n and λ_n be as in the definition of T and put $r_n = \lambda_n(x_n - x_\circ) - x$, so that $x_n - x_\circ = (x + r_n)/\lambda_n$. Since $x \neq 0$, as $n \to \infty$ we have that $r_n \to 0$ and $|x + r_n|/|x| \to 1$. Furthermore we have from the definition of ∇ that as $n \to \infty$

$$\frac{|g_a(x_n) - \nabla g_a(x_\circ)(x_n - x_\circ)|}{|x_n - x_\circ|} = \frac{|g_a(x_n) - g_a(x_\circ) - \nabla g_a(x_\circ)(x_n - x_\circ)|}{|x_n - x_\circ|} \to 0,$$

and hence, replacing $x_n - x_\circ$ by $(x + r_n)/\lambda_n$ and using $|x + r_n|/|x| \to 1$, it follows that $\lambda_n g_a(x_n) - \nabla g_a(x_\circ) r_n \to \nabla g_a(x_\circ)x$. But $x_n \in G_a$ so that $g_a(x_n) \leqslant 0$ and $r_n \to 0$; hence $\nabla g_a(x_\circ)x \leqslant 0$ and so $x \in S$.

If we introduce the dual cones

$$(5.2.13) \qquad T^* = \{x^* : x^*x \geqslant 0 \text{ for all } x \in T\} \quad \text{and} \quad S^* = \{x^* : x^*x \geqslant 0 \text{ for all } x \in S\}$$

(in the literature the signs are often reversed here and then one speaks of the *polar* cones) it immediately follows from this lemma that

$$(5.2.14) \qquad S^* \subset T^*.$$

The next lemma gives another formulation for S^*.

5.2.15 **Lemma.** $S^* = \{x^* : x^* + y_a^* \nabla g_a(x_\circ) = 0 \text{ for some } y_a^* \geqslant 0\}$.

Proof. If $x^* \in S^*$ take $c = x^*$ and $A = \nabla g_a(x_\circ)$ in Lemma 5.2.3. Then the existence of the required y_a^* immediately follows.

Conversely if $x^* + y_a^* \nabla g_a(x_\circ) = 0$ for some $y_a^* \geqslant 0$ then, of course, $x^* \in S^*$.

We are now ready to formulate the constraint qualification we had in mind. It is simply that

$$(5.2.16) \qquad S^* \supset T^*, \quad \text{or, equivalently,} \quad S^* = T^*.$$

Notice that the definition of T^* is entirely free from differentials, whereas S^* is defined in terms of $\nabla g_a(x_\circ)$. Hence (5.2.16) says that S^* can be constructed without using differentials explicitly. Also notice that (5.2.16) does not imply in general that $S = T$, because T

is not necessarily convex. By changing from T to T^* we implicitly perform a certain 'convexification'.

From a computational point of view we can improve condition (5.2.16). Divide the active constraints into two (or more) subsets and define sets like S for each of them. Call these S_1 and S_2 and include the corresponding constraints as $g_{a,1}$ and $g_{a,2}$ respectively. Then $S = S_1 \cap S_2$ and $S^* = S_1^* + S_2^*$. This follows from Lemma 5.2.15, because if $x^* \in S^*$ then $x^* = -y_1^* \nabla g_{a,1}(x_o) - y_2^* \nabla g_{a,2}(x_o)$ for some $y_1^* \geqslant 0, y_2^* \geqslant 0$, hence $x^* \in S_1^* + S_2^*$ and so $S^* \subset S_1^* + S_2^*$. The reverse inclusion follows in a similar way. Instead of (5.2.16) we may now write

(5.2.17) $$S_1^* + S_2^* \supset T^*,$$

which has the advantage that the dual cones of S_i may be computed separately.

Now we prove that (5.2.16) is indeed sufficient to ensure the existence of the desired y_o^*.

5.2.18 **Theorem.** Let f, g and x_o be as in Theorem 5.2.8. Then the Kuhn–Tucker conditions (5.2.1) hold if $S^* \supset T^*$.

Proof. In view of Lemma 5.2.15 all we have to show is that $\nabla f(x_o) \in T^*$, i.e. that $\nabla f(x_o)x \geqslant 0$ for all $x \in T$. Let $x \in T$, $x \neq 0$ and let λ_n and x_n be as in the definition of T. Again let $r_n = \lambda_n(x_n - x_o) - x$. Since $g_a(x_n) \leqslant 0$, since $g_i(x_n) < 0$ if g_i is inactive and n is large enough, and since x_o is optimal, it follows that $f(x_n) \geqslant f(x_o)$ if n is large enough.

On the other hand, applying to f the same reasoning as we applied to g_a in the proof of Lemma 5.2.12, we have that, as $n \to \infty$,

$$\lambda_n f(x_n) - \lambda_n f(x_o) - \nabla f(x_o) r_n \to \nabla f(x_o)x.$$

Since $\lambda_n f(x_n) \geqslant \lambda_n f(x_o)$ and $r_n \to 0$ it follows from this that indeed $\nabla f(x_o)x \geqslant 0$.

Let us now reverse the reasoning and not try to derive *necessary* conditions resulting from the solvability of the given problem, but specify *sufficient* conditions for solvability. In the next theorem this is done by means of convexity. As might be expected the result is *global* solvability. In the next section, however, we shall not rely on convexity, but introduce second-order derivatives and specify conditions in terms of them. Then we can only establish *local* solvability.

5.2.19 **Theorem.** Let f and g be as before and let x_0 be such that the Kuhn–Tucker conditions hold. Then x_0 is a solution of the problem of finding $\inf\{f(x): g(x) \leqslant 0\}$ if f and g are convex.

Proof. Let x be such that $g(x) \leqslant 0$. Then by Lemma 5.1.10

$$f(x)-f(x_0) \geqslant f(x)-f(x_0)+y_0^* g(x)-y_0^* g(x_0)$$
$$\geqslant \nabla(f(x_0)+y_0^* g(x_0))(x-x_0) = 0,$$

that is if x is feasible then $f(x) \geqslant f(x_0)$, hence x_0 is optimal.

So far we have been considering cases where both X and Y are *finite*-dimensional. If we try to replace Y by an *infinite*-dimensional space we run into trouble, for two reasons. First of all because the required generalization of Lemma 5.2.3 does not hold; and secondly because it is then not so easy to distinguish active constraints from inactive constraints. To see this recall the simple example of 3.11 where $x \leqslant 1/i$, $i = 1, 2, \ldots$ and $x_0 = 0$. Each separate constraint $x \leqslant 1/i$ is inactive, but together they are active because they imply the active constraint $x \leqslant 0$.

On the other hand in Lemma 5.2.3 we are allowed to replace R_n by any Banach space X, c by an element of X^* and A by a continuous linear mapping from X to R_m. Given this, it is not difficult to verify that all the results of this section up to now can be generalized if we replace ordinary differentiability by Fréchet differentiability:

5.2.20 **Theorem.** Let X be a Banach space and let $f: X \to R$ and $g: X \to Y = R_m$ be Fréchet differentiable. Let g_a, G_a and T be as before and let

(5.2.21) $S = \{x: g_a'(x_0)x \leqslant 0\}.$

(*a*) If x_0 is a solution of the problem of finding $\inf\{f(x): g(x) \leqslant 0\}$ then

   (*aa*) a necessary and sufficient condition for the existence of a y_0^* such that

(5.2.22) $f'(x_o)+y_o^*g'(x_o)=0, \quad y_o^*g(x_o)=0, \quad y_o^* \geqslant 0,$

is that for all $x \in X$

(5.2.23) $g_a'(x_o)x \leqslant 0$ implies that $f'(x_o)x \geqslant 0,$

(*ab*) (5.2.22) holds for some y_o^* if Slater's condition (5.2.7) is satisfied, and

(*ac*) (5.2.22) holds for some y_o^* if $S^* \supset T^*$.

(*b*) Conversely if (5.2.22) is true for some y_o^* and some x_o such that $g(x_o) \leqslant 0$, then $f(x_o) = \inf\{f(x): g(x) \leqslant 0\}$, if f and g are convex. With the exception of part (*ab*), *equality constraints* are permitted, for example by replacing them by two inequality constraints.

5.3 Finitely many constraints; second-order conditions

From elementary analysis we know that the second derivative of a function is nonnegative at a local minimum, and conversely that if the first derivative is zero and the second derivative is positive at some point then we have a local minimum there. The purpose of this section is to generalize these facts for our more general problems. This requires the introduction of second-order Fréchet derivatives. As we have remarked already, when $q''(x_o)$ is the second-order Fréchet derivative at x_o of $q: X \to \tilde{X}$, then $q''(x_o) \in L(X, L(X, \tilde{X}))$, so that $q''(x_o)\,x_1 \in L(X, \tilde{X})$ and $q''(x_o)\,x_1, x_2 \in \tilde{X}$. In particular, if $x = R$ then $q''(x_o)x_1x_2$ is a quadratic form, and instead of the nonnegativity or the positivity of the second derivative of an elementary function we should talk of $q''(x_o)$ being positive semi-definite or positive definite, which means that $q''(x_o)xx \geqslant 0$ for all x, or $q''\{x_o)xx > 0$ for all $x \neq 0$, respectively.

If for some solution x_o of the problem of finding $\inf\{f(x): g(x) \leqslant 0\}$ and some y_o^* we have that $L'(x_o, y_o^*) = f'(x_o)+y_o^*g'(x_o) = 0$ then we should not expect that $L''(x_o, y_o^*)$ is positive semi-definite. All we can hope for is that $L''(x_o, y_o^*)xx \geqslant 0$ if x is tangent to the active constraints.

5.3.1 **Example.** Let $X = Y = R_2$ and

$$\alpha = \inf\{x^t\begin{pmatrix}1 & 0\\ 0 & -1\end{pmatrix}x : (1, 2)x \leqslant 0, \quad -(1,2)x \leqslant 0\},$$

so that in fact we have the equality constraint $(1,2)x = 0$. Then $x_0 = 0$ and $L''(x_0, y_0^*)\,xx = x^t\begin{pmatrix}2 & 0\\ 0 & -2\end{pmatrix}x$, which can take negative values, unless we require that $(1,2)x = 0$.

5.3.2 **Theorem.** Let X be a Banach space and $f: X \to R$, $g: X \to Y = R_m$. Let x_0 be a solution of the problem of finding $\inf\{f(x): g(x) \leqslant 0\}$. Let f and g be twice Fréchet differentiable in a neighbourhood of x_0 and let f'' and g'' be continuous at x_0. Assume further that $g_a'(x_0)$ is a mapping onto Y. Finally assume that the first-order conditions at x_0 are satisfied, hence that (5.2.22) is true for some y_0^*. Then with $L(x, y^*) = f(x) + y^* g(x)$ we have that

$$L''(x_0, y_0^*)x_1 x_1 \geqslant 0 \quad \text{if} \quad g_a'(x_0)x_1 = 0. \tag{5.3.3}$$

Proof. Let x_1 be such that $g_a'(x_0)x_1 = 0$. By Lemma 5.1.7 with $q = g_a$ and x_0 and x_1 as here, we find that $g_a(x_0 + \lambda x_1 + o(\lambda)) = 0$. Set $x = \lambda x_1 + o(\lambda)$. Since $x_0 + x$ is feasible if λ is small enough, $f(x_0 + x) \geqslant f(x_0)$ for such λ. From Lemma 5.1.6(*d*) we have for some number θ between 0 and 1 that

$$f(x_0+x) + y_0^* g(x_0+x) - f(x_0) - y_0^* g(x_0) - f'(x_0)\,x - y_0^* g'(x_0)x$$
$$= \tfrac{1}{2} f''(x_0 + \theta x)xx + \tfrac{1}{2} y_0^* g''(x_0 + \theta x)xx.$$

Combining this with $f(x_0+x) \geqslant f(x_0)$, $g_a(x_0+x) = 0$ and the first-order conditions we obtain $L''(x_0 + \theta x)xx \geqslant 0$. Substituting $\lambda x_1 + o(\lambda)$ for x, dividing by λ^2 and letting λ tend to zero we arrive at the desired result, since f'' and g'' are assumed to be continuous at x_0.

Equality constraints can be handled by this theorem. Further we can generalize the theorem to the case where Y is any Banach space but $g(x) \leqslant 0$ is replaced by $g(x) = 0$. The restriction to equality constraints is necessary because of the difficulty of saying which constraints are active, and the generalization to a Banach space for Y is possible because Lemma 5.1.7 is general enough and because we have assumed that the first-order conditions hold.

Let us now consider *sufficient second-order conditions.*

5.3.4 **Theorem.** Let X, Y, f and g be as in Theorem 5.3.2, and assume that X is *reflexive*. Let the first-order conditions (5.2.22) be satisfied for some y_0^* and some x_o such that $g(x_o) \leqslant 0$. Let f and g be twice Fréchet differentiable in a neighbourhood of x_o and let f'' and g'' be continuous at x_o. Let y_a^* be the vector consisting of all positive components of y_0^* and let g_a be the corresponding part of g (this is not to say that g_a is the active part of g, for g_i may be active at x_o while the ith component of y_0^* is zero). Finally assume that $L''(x_o, y_0^*)\, x_1 x_1 > 0$ if $g_a'(x_o)x_1 = 0$ and $x_1 \neq 0$, that is assume that $L''(x_o, y_0^*)$ is positive definite on the subspace $\{x_1 \colon g_a'(x_o)x_1 = 0\}$. Then x_o is a *strict* local minimum of the problem of finding $\inf\{f(x) \colon g(x) \leqslant 0\}$ which means that if $g(x_o + x) \leqslant 0$ and $x \neq 0$ then $f(x_o + x) > f(x_o)$ if x is small enough.

Proof. To the contrary suppose x_o is not a strict local minimum. Then for some sequence $x^i \to 0$, $x^i \neq 0$, $f(x_o + x^i) \leqslant f(x_o)$ while $g(x_o + x^i) \leqslant 0$. Because X is reflexive its unit ball is compact, hence $x^i/|x^i| \to x_1$ for some x_1 and some subsequence, and $|x_1| = 1$. Since $f(x_o + x^i) \leqslant f(x_o)$ we must have that $f'(x_o)x_1 \leqslant 0$ and since $g(x_o + x^i) \leqslant 0$ and $y_0^* g(x_o) = 0$ that $y_0^* g'(x_o)x_1 \leqslant 0$. From this and the first-order conditions it follows that $f'(x_o)x_1 = y_0^* g'(x_o)x_1 = 0$ so that $g_a'(x_o)x_1 = 0$. By assumption then $L''(x_o, y_0^*)x_1 x_1 > 0$.

But for some θ^i between 0 and 1 we have again that

$$L(x_o + x^i, y_0^*) - L(x_o, y_0^*) = \tfrac{1}{2}L''(x_o + \theta^i x^i, y_0^*)x^i x^i$$

and since $f(x_o + x^i) \leqslant f(x_o)$ and $g(x_o + x^i) \leqslant 0$ it follows from this if we divide by $|x^i|^2$ that $L''(x_o, y_0^*)x_1 x_1 \leqslant 0$ which is a contradiction.

5.4 Problems with arbitrary constraints; first-order conditions; linearizing sets

This section is concerned with generalizations of the results of the previous two, in two respects. First of all Y is no longer required to be finite-dimensional and secondly the set C will be given a more essential part to play. The latter implies that we are not aiming for the Kuhn–Tucker conditions but rather for the 'minimum principle' type of conditions like (5.1.4).

The first theorem below is not suitable for problems with equality constraints because the regularity condition there is of the 'Slater type', and the second one is concerned with equality constraints only (except for the constraint $x \in C$). Theorems for problems with mixed constraints can be obtained from these two. Having done something similar in 3.13 (see Theorem 3.13.2), however, we shall leave this to the reader.

5.4.1 **Theorem.** Let X and Y be Banach spaces, Y with a positive cone with nonempty interior; and let $f: X \to R$, $g: X \to Y$ and $C \subset X$, with C convex. Let x_o be a solution of the problem of finding $\inf\{f(x): g(x) \leqslant 0, x \in C\}$. Assume that f and g are Fréchet differentiable at x_o and that for some $\hat{x} \in C$, $g(x_o)+g'(x_o)(\hat{x}-x_o) < 0$. Then there exists a $y_0^* \in Y^*$, the dual of Y, such that

$$(5.4.2) \qquad (f'(x_o)+y_0^* g'(x_o))(x-x_o) \geqslant 0 \qquad \text{if} \quad x \in C, \quad y_0^* g(x_o) = 0, \quad y_0^* \geqslant 0.$$

Proof. Consider the problem of finding

$$\tilde{\alpha} = \inf\{f'(x_o)(x-x_o): g(x_o)+g'(x_o)(x-x_o) \leqslant 0, \quad x \in C\}.$$

This is a convex optimization problem, and in fact is a *linearized* version of the given one. Since $x_o \in C$, obviously $\tilde{\alpha} \leqslant 0$. And in fact $\tilde{\alpha} = 0$ because otherwise there would exist an $x \in C$ such that $f'(x_o)(x-x_o) < 0$ and $g(x_o)+g'(x_o)(x-x_o) \leqslant 0$, hence

$$f'(x_o)(\lambda(\hat{x}-x_o)+(1-\lambda)(x-x_o)) < 0$$

and $$g(x_o)+g'(x_o)(\lambda(\hat{x}-x_o)+(1-\lambda)(x-x_o)) < 0$$

for $\lambda > 0$ and small enough. Hence by Lemma 5.1.6(*b*), (*c*), setting

$$\tilde{x} = x_o + \tau\lambda(\hat{x} - x_o) + \tau(1-\lambda)(x - x_o) = (1-\tau)x_o + \tau\lambda\hat{x} + \tau(1-\lambda)x,$$

we see that $f(\tilde{x}) < f(x_o)$ and $g(\tilde{x}) < 0$ for $\tau > 0$ and small enough. But $\tilde{x}$ is a convex combination of x_o, $\hat{x}$ and x, all belonging to C, so that $\tilde{x} \in C$. It follows that $\tilde{x}$ is feasible and that $f(\tilde{x}) < f(x_o)$ which cannot be.

We can now invoke results about Lagrangian duality for problems with inequality constraints, such as Theorem 3.11.2 and Corollary 3.11.5, to obtain the existence of a $y_0^* \geqslant 0$ such that $y_0^* g(x_o) = 0$ and such that

$$\inf\{f'(x_o) + y_0^* g'(x_o))(x - x_o) \colon x \in C\} = 0,$$

which is just (5.4.2).

5.4.3 **Theorem.** Let X and Y be Banach spaces, $f\colon X \to R$, $h\colon X \to Y$ and C a convex subset of X. Let x_o be a solution of the problem of finding $\inf\{f(x)\colon h(x) = 0, x \in C\}$. Assume that f is Fréchet differentiable at x_o, that h is Fréchet differentiable in a neighbourhood of x_o, and that h' is continuous at x_o. Further assume that for some $\hat{x} \in \operatorname{ri} C$, $h'(x_o)(\hat{x} - x_o) = 0$. Finally assume that $h'(x_o)$ is a mapping onto Y. Then there exists a $y_0^* \in Y^*$, the dual of Y, such that

$$(f'(x_o) + y_0^* h'(x_o))(x - x_o) \geqslant 0 \quad \text{if} \quad x \in C. \tag{5.4.4}$$

Proof. Again consider a *linearized* version of the given problem, that is let

$$\tilde{\alpha} = \inf\{f'(x_o)(x - x_o) \colon h'(x_o)(x - x_o) = 0, \quad x \in C\}.$$

Again $\tilde{\alpha} \leqslant 0$, and in fact $\tilde{\alpha} = 0$. Suppose the latter were not true. Then for some $x \in C$, $f'(x_o)(x - x_o) < 0$ and $h'(x_o)(x - x_o) = 0$. Let $\lambda > 0$ be so small that

$$f'(x_o)(\lambda(\hat{x} - x_o) + (1-\lambda)(x - x_o)) < -\epsilon$$

for some $\epsilon > 0$. Since

$$h'(x_o)(\lambda(\hat{x} - x_o) + (1-\lambda)(x - x_o)) = 0$$

it follows from Lemma 5.1.7 that

$$h(x_0+\tau\lambda(\hat{x}-x_0)+\tau(1-\lambda)(x-x_0)+o(\tau))=0.$$

Assume for the time being that $\operatorname{ri} C=\operatorname{int} C$. Then since $\hat{x}\in\operatorname{int} C$, $\hat{x}+o(\tau)/\tau\lambda\in C$ if τ is small enough. Take $\tau>0$ and such that $f'(x_0)o(\tau)/\tau<\epsilon$. Then

$$\tilde{x}=x_0+\tau\lambda(\hat{x}-x_0)+\tau(1-\lambda)(x-x_0)+o(\tau)$$

is a convex combination of x_0, $\hat{x}+o(\tau)/\tau\lambda$ and x, all belonging to C. Hence $\tilde{x}\in C$ and $h(\tilde{x})=0$, but

$$f'(x_0)(\lambda(\hat{x}-x_0)+(1-\lambda)(x-x_0)+o(\tau)/\tau)<0,$$

hence by Lemma 5.1.6(*b*) $f(\tilde{x})<f(x_0)$ if τ is small enough, a contradiction. It follows that $\tilde{\alpha}=0$.

By Theorem 3.12.1 the desired result now follows immediately, at least if $\operatorname{ri} C=\operatorname{int} C$. If this is not true then we can replace X by $L(C)$, which as we have seen before is itself a Banach space.

Stating and proving *sufficiency* theorems in the spirit of Theorem 5.2.19 seems fairly obvious and is left to the reader.

One aspect of what we have been doing so far in this section is not entirely satisfactory, namely that we have linearized the functions f, g and h, but have left the set C untouched. *How do we linearize C*? Constructing a first-order approximation of a function means disregarding terms of order higher than one. So if $x_0\in C$ a natural requirement for a linearization K of C at x_0 seems to be that

(5.4.5) if $x\in K$ then $x_0+\tau x+o(\tau)\in C$, for some $o(\tau)$.

(A weaker requirement is to replace the continuously varying τ in (5.4.5) by a discretely varying τ; then the cone of tangents of C at x_0, defined in (5.2.9), would be an example of a linearization of C at x_0.) From this it follows immediately that K must be a *cone with apex at the origin*, so that not K but x_0+K really approximates C at x_0. The following cones all satisfy condition (5.4.5) if $x_0\in C$:

(*a*) a cone that is closely related to the cone of tangents of C at x_0 (see 5.2.9), i.e.

(5.4.6) $$\tilde{T} = \{x: x = \lim_{\lambda\to\infty} \lambda(x(\lambda)-x_0) \text{ such that } x(\lambda) \in C \text{ and } \lim_{\lambda\to\infty} x(\lambda) = x_0\};$$

(*b*) the so-called *cone of internal directions of C at* x_0, i.e.

(5.4.7)
$$D = \{x: x_0+\tau(x+U) \subset C \quad \text{for some neighbourhood } U \text{ of the origin and all } \tau,\ \ 0<\tau\leqslant\tau_0 \quad \text{for some } \tau_0>0\};$$

(*c*) the cone

(5.4.8) $$\{x: x_0+\tau x \in C \text{ for all } \tau,\ 0\leqslant\tau\leqslant\tau_0 \quad \text{for some} \quad \tau_0>0\};$$

and so on.

5.4.9 **Definition.** We say that K is a *first-order approximation of C at* x_0 if $x \in K$ implies that $x_0+\tau x+o(\tau)\in C$, for some $o(\tau)$.

We can now state and prove two theorems which in all respects are 'local'.

5.4.10 **Theorem.** Let X and Y be Banach spaces, Y with a positive cone with nonempty interior, and let $f: X\to R$, $g: X\to Y$ and $C\subset X$. Let x_0 be a solution of the problem to find $\inf\{f(x): g(x)\leqslant 0,\ x\in C\}$ and let K be a first-order approximation of C at x_0. Assume that K is convex, that f and g are Fréchet differentiable at x_0 and that for some $\hat{x}\in K$, $g(x_0)+g'(x_0)\hat{x}<0$ (which is analogous to Slater's constraint qualification (5.2.7)). Then there exists a $y_0^*\in Y^*$, the dual of Y, such that

(5.4.11) $$(f'(x_0)+y_0^*g'(x_0))x \geqslant 0 \quad \text{if} \quad x\in K,\quad y_0^*g(x_0)=0,\quad y_0^*\geqslant 0;$$

hence $f'(x_0)+y_0^*g'(x_0)\in K^*$, the dual cone of K.

Proof. Let

$$\tilde{\alpha} = \inf\{f'(x_0)x: g(x_0)+g'(x_0)x\leqslant 0,\quad x\in K\}.$$

Since $K \neq \varnothing$, $0\in \operatorname{cl} K$ so that $\tilde{\alpha}\leqslant 0$. And again we can show that $\tilde{\alpha}=0$, for if this is not true then, as before,

$$f'(x_0)(\lambda\hat{x}+(1-\lambda)x) < 0$$

and
$$g(x_0)+g'(x_0)(\lambda\hat{x}+(1-\lambda)x) < 0$$

for some $x \in K$ and some λ, $0 < \lambda < 1$. Because $\hat{x} \in K$, $x \in K$ and K is convex, $\lambda\hat{x}+(1-\lambda)x \in K$, hence

$$\tilde{x} = x_0+\tau\lambda\hat{x}+\tau(1-\lambda)x+o(\tau) \in C.$$

Take τ so small that

$$f'(\lambda\hat{x}+(1-\lambda)x+o(\tau)/\tau) < 0$$

and
$$g(x_0)+g'(x_0)(\lambda\hat{x}+(1-\lambda)x+o(\tau)/\tau) < 0.$$

By Lemma 5.1.6(*b*), (*c*) we have again that $f(\tilde{x}) < f(x_0)$ and that $g(\tilde{x}) < 0$ if τ is small enough. But $\tilde{x} \in C$, so that we have arrived at a contradiction.

The remainder of the proof now proceeds as before.

Next we consider *equality* constraints. Now we must require a little more of the cone approximating C.

5.4.12 **Theorem.** Let X and Y be Banach spaces, $f\colon X \to R$, $g\colon X \to Y$ and $C \subset X$. Let x_0 be a solution of the problem of finding $\inf\{f(x)\colon h(x) = 0,\ x \in C\}$, and let D be the cone of interval directions of C at x_0, defined by (5.4.7). Assume that D is convex, f is Fréchet differentiable at x_0, h is Fréchet differentiable in a neighbourhood of x_0 and h' is continuous at x_0. Further assume that for some $\hat{x} \in \operatorname{ri} D$, $h'(x_0)\hat{x} = 0$ and that $h'(x_0)$ is a mapping onto Y. Then there exists a $y_0^* \in Y^*$, the dual of Y, such that

(5.4.13)
$$(f'(x_0)+y_0^*h'(x_0)x \geqslant 0 \quad \text{if} \quad x \in D.$$

Proof. The proof is similar to previous proofs. Let

$$\tilde{\alpha} = \inf\{f'(x_0)x\colon h'(x_0)x = 0, \quad x \in D\}.$$

If $\tilde{\alpha} < 0$, then we find that

$$h(x_0+\tau\lambda\hat{x}+\tau(1-\lambda)\ x+o(\tau)) = 0$$

and
$$f(x_0+\tau\lambda\hat{x}+\tau(1-\lambda)x+o(\tau)) < f(x_0)$$

for some $x \in D$, some λ, $0 < \lambda < 1$, and some $o(\tau)$. Since D is convex, $\lambda\hat{x}+(1-\lambda)x \in D$, hence by the definition of D,

$$x_o + \tau\lambda\hat{x} + \tau(1-\lambda)x + \tau U \subset C$$

for some neighbourhood U of the origin, and τ small. Take τ so small that $o(\tau)/\tau \in U$; then

$$x_o + \tau\lambda\hat{x} + \tau(1-\lambda)x + o(\tau) \in C$$

and again we have arrived at a contradiction.

In the next theorem we consider *inequality* constraints, but these may be replaced by *equality* constraints because we do not require that the interior of the relevant positive cone is nonempty. Its proof is based on *strong separation* in X^* and that is why we must assume that X is *reflexive*. The theorem itself is analogous to Theorem 5.2.4 involving the condition that $\nabla g_a(x_o)x \leqslant 0$ should imply $\nabla f(x_o)x \geqslant 0$, so that in some way or other we must deal with active and inactive constraints. This is done implicitly as follows. Given $x_o \in C$ let

(5.4.14) $K(x_o) = \{x\colon x_o + \lambda x \in C \text{ for some } \lambda > 0\} \cup \{0\}$

which is the *cone generated by C at* x_o (see Definition 3.8.5). Notice, incidentally, that $K(x_o)$ in general does not satisfy condition (5.4.5), hence that in general it cannot serve as a first-order approximation of C at x_o. Consider the following implication,

(5.4.15)

$$\left.\begin{array}{l} x \in \operatorname{cl} K(x_o) \quad \text{and} \quad y^*g'(x_o)x \leqslant 0 \\ \text{for all } y^* \geqslant 0 \text{ such that } y^*g(x_o) = 0 \end{array}\right\} \text{ implies that } f'(x_o)x \geqslant 0.$$

If $C = K(x_o) = X$ and $Y = R_m$ this reduces to

(5.4.16) $\qquad g_a'(x_o)x \leqslant 0 \quad \text{implies that} \quad f'(x_o)x \geqslant 0,$

which is what we had before.

Moreover we must require that

(5.4.17)

$$G^* = \{x^*\colon \text{for some } y^* \geqslant 0 \text{ such that } y^*g(x_o) = 0, \\ (x^* + y^*g'(x_o))x \geqslant 0 \quad \text{if} \quad x \in K(x_o)\}$$

is a *closed* set. To see the meaning of G^* take again $C = X$ and $Y = R_m$. Then $G^* = \{x^*: x^* + y_a^* g_a'(x_o) = 0 \text{ for some } y_a^* \geqslant 0\}$. From Appendix D we know that this set is closed.

5.4.18 **Theorem.** Let X and Y be Banach spaces, Y with a positive cone (which may degenerate to $\{0\}$) and let $C \subset X$. Let x_o be a solution of the problem of finding $\inf\{f(x): g(x) \leqslant 0, x \in C\}$, and let $f: X \to R$ and $g: X \to Y$ be Fréchet differentiable at x_o. Then for some $y_o^* \in Y^*$, the dual of Y,

(5.4.19) $$(f'(x_o) + y_o^* g'(x_o))x \geqslant 0$$
$$\text{if} \quad x \in K(x_o), \quad y_o^* g(x_o) = 0, \quad y_o^* \geqslant 0$$

if X is reflexive, $K(x_o)$ is convex, the implication (5.4.15) is true, and G^* is closed.

Proof. Put $c = f'(x_o)$ and $A = g'(x_o)$. All we have to show is that $c \in G^*$. If this is not so then we can separate c strongly from G^*, because G^* is convex and closed. This is a separation in X^*, but since X is reflexive it follows that for some $x' \in X$ and some $\tau > 0$, $x^*x' \geqslant cx' + \tau$ if $x^* \in G^*$. Because $0 \in G^*$ we have that $cx' + \tau \leqslant 0$ or $cx' < 0$, and because $-y^*A \in G^*$ if $y^* \geqslant 0$ and $y^*g(x_o) = 0$ that $-y^*Ax' \geqslant cx' + \tau$, so that $y^*Ax' \leqslant 0$ for such y^*. Since $cx' < 0$ it follows from this and (5.4.15) that $x' \notin \text{cl}\, K(x_o)$ which is convex because $K(x_o)$ is. Hence once again we can apply strong separation, this time between x' and $\text{cl}\, K$, and so for some x'^* and some $\tau' > 0$, $x'^*x \geqslant x'^*x' + \tau'$ if $x \in \text{cl}\, K(x_o)$. Because $0 \in \text{cl}\, K(x_o)$ we have $x'^*x' + \tau' \leqslant 0$, and hence $x'^*x' < 0$.

On the other hand, since $K(x_o)$ is a cone, $x'^*\lambda x \geqslant x'^*x' + \tau'$ if $x \in K(x_o)$ and $\lambda > 0$, so that $x'^*x \geqslant 0$ if $x \in K(x_o)$; but this means that $x'^* \in G^*$ (with $y^* = 0$). From the first separation it now follows that $x'^*x' \geqslant cx' + \tau$. Here we may replace x'^* by $\lambda x'^*$ if $\lambda > 0$, so that $x'^*x' \geqslant 0$, which contradicts $x'^*x' < 0$.

5.5 Problems with arbitrary constraints; second-order conditions

In this section we continue the discussion of 5.4, but although we allow constraints of the form $x \in C$, we shall require that in constraints of the form $g(x) \leqslant 0$ the vector $g(x)$ is finite-dimensional. This is because we want to distinguish active constraints from inactive constraints, even though in principle it is possible to do so implicitly when $g(x)$ is infinite-dimensional by proceeding as in Theorem 5.4.18.

We have again a necessity theorem and a sufficiency theorem. In the former we shall use the cone of internal directions as a first-order approximation of C at x_0, whereas in the latter we shall replace this cone by the cone of tangents of C at x_0. The reason for this becomes clear if we look at the proof of Theorem 5.3.4, where we constructed a sequence $x^i \to 0$, $x^i \neq 0$ such that $x^i/|x^i| \to x_1$ for some x_1. If we know that $x_0 + x^i \in C$, then it easily follows that x_1 is an element of the cone of tangents.

Both theorems are formulated with *inequality* constraints, but these may be replaced by *equality* constraints.

5.5.1 **Theorem** (necessary second-order conditions). Let X be a Banach space, $f: X \to R$, $g: X \to Y = R_m$ and $C \subset X$. Let x_0 be a solution of the problem of finding $\inf\{f(x): g(x) \leqslant 0, x \in C\}$. Let f and g be twice Fréchet differentiable at x_0 and let f'' and g'' be continuous at x_0. Further assume that $g_a'(x_0)$ is a mapping onto Y. Finally assume that the first-order conditions with respect to D, the cone of internal directions, are satisfied:

$$(f'(x_0) + y_0^* g'(x_0))x \geqslant 0 \quad \text{if} \quad x \in D, \quad y_0^* g(x_0) = 0, \quad y_0^* \geqslant 0. \tag{5.5.2}$$

Then

$$L''(x_0, y_0^*)x_1 x_1 \geqslant 0 \quad \text{if} \quad g_a'(x_0)x_1 = 0 \quad \text{and} \quad x_1 \in D. \tag{5.5.3}$$

Proof. The wording of the proof can be taken as identical to that of the proof of Theorem 5.3.2, but '$x_0 + x$ feasible' now means that x must also satisfy the condition that $x_0 + x \in C$. This is, of course, true if we take in $x = \lambda x_1 + o(\lambda)$, λ positive and small enough, for $x_1 \in D$ implies that $x_0 + \lambda x_1 + \lambda U \subset C$ for some neighbourhood U of the origin and all λ such that for some $\lambda_0 > 0$, $0 < \lambda \leqslant \lambda_0$.

5.5.4 **Theorem** (sufficient second-order conditions). Let X, Y, C, f and g be as in Theorem 5.5.1, and assume that X is *reflexive.*

Let the first-order conditions (5.5.2) be satisfied but with D replaced by T, the cone of tangents of C at x_o, and let $g(x_o) \leqslant 0$ and $x_o \in C$. Let f and g be twice Fréchet differentiable in a neighbourhood of x_o and let f'' and g'' be continuous at x_o. Let y_a^* consist of all positive components of y_o^* and let g_a be the corresponding part of g (notice the comment in Theorem 5.3.4). Finally assume that $L''(x_o, y_o^*)x_1 x_1 > 0$ if $g_a'(x_o)x_1 = 0$ and if $x_1 \in T$. Then x_o is a *strict* local minimum of the problem of finding $\inf\{f(x): g(x) \leqslant 0, x \in C\}$.

Proof. The proof is almost identical to that of Theorem 5.3.4. The x^i now also satisfy the condition that $x_o + x^i \in C$. Taking in (5.4.6) $\lambda = \lambda_i = 1/|x^i|$ it follows that $x_1 \in T$, and we can conclude that if the theorem were not true we would have $L''(x_o, y_o^*)x_1 x_1 \leqslant 0$.

6
Some other approaches

6.1 The fixed point approach

We continue here the discussion of 4.5, where we considered the Lagrangian L a competitor of the bifunction F, in that everything can be derived not only from F, but also from L, at least if the transformation from F to L, i.e.

(6.1.1) $$L(x, y^*) = \inf_Y (F(x, y) + y^* y)$$

may be inverted to

(6.1.2) $$F(x, y) = \sup_{Y^*} (L(x, y^*) - y^* y).$$

Conversely, we could start from L, define F by (6.1.2) and everything would be fine if we were allowed to invert (6.1.2) back to (6.1.1). It is not difficult to verify that the latter is equivalent to L being equal to $-(-L)^{**}$ where conjugation is taken with respect to y^* and y only. Thus if $L = -(-L)^{**}$ then instead of starting from F we can equally well start from L. But even if this is not so there is a point in starting from L if we are willing to define the primal problem as that of finding $\inf_X \sup_{Y^*} L(x, y^*)$ and, as before, let the dual problem be that of finding $\sup_{Y^*} \inf_X L(x, y^*)$. Still another possibility is that the problem we are given is to find a saddle-point of L, so that we are not primarily interested in primal and dual problems at all. An example of this is where we are given a game, such as the one described in Example 1.2.7.

In any case let a function $L: X \times Z \to [-\infty, +\infty]$ be given, where $Z = Y^*$. Or, avoiding infinite function values, let S be a subset of $X \times Z$ and let $L: S \to R$. Assume that S is 'rectangular', that is that $S = C \times D$, with $C \subset X$ and $D \subset Z$. The point (x_0, z_0) is a saddle-point of L (with respect to S) if

(6.1.3) $$x_0 \in C, \quad z_0 \in D, \quad L(x_0, z) \leqslant L(x_0, z_0) \leqslant L(x, z_0) \quad \text{for all } x \in C \text{ and all } z \in D.$$

An equivalent definition is that $x_0 \in C$, $z_0 \in D$ and that

$$(6.1.4) \qquad \begin{aligned} x_0 &\in \{x' : L(x', z_0) = \min_C L(x, z_0)\}, \\ z_0 &\in \{z' : L(x_0, z') = \max_D L(x_0, z)\} \end{aligned}$$

or, defining the sets $A(z)$ and $B(x)$, with $z \in D$ and $x \in C$, by

$$(6.1.5) \qquad \begin{aligned} A(z) &= \{x' : L(x', z) = \min_C L(x, z)\}, \\ B(x) &= \{z' : L(x, z') = \max_D L(x, z)\}, \end{aligned}$$

that

$$(6.1.6) \qquad (x_0, z_0) \in A(z_0) \times B(x_0).$$

Another reformulation is obtained if we define the following *point-to-set* mapping from $C \times D$ to the class of all subsets of $C \times D$ (hence to the *power set* of $C \times D$):

$$(6.1.7) \qquad T(x, z) = A(z) \times B(x), \quad x \in C, \quad z \in D.$$

Terming a point $(x, z) \in C \times D$ a *fixed point* of T whenever $(x, z) \in T(x, z)$, we can now say that (x_0, z_0) is a *saddle-point of L if and only if* (x_0, z_0) *is a fixed point of T.*

An interesting generalization of this relationship is where L is replaced by two functions ϕ_1 and ϕ_2, and where the problem is to find (x_0, z_0) such that

$$(6.1.8) \qquad \begin{aligned} & x_0 \in C, \quad z_0 \in D, \\ & \phi_1(x_0, z_0) = \min_C \phi_1(x, z_0), \\ & \phi_2(x_0, z_0) = \min_D \phi_2(x_0, z). \end{aligned}$$

In terms of game theory this means that we are given *two* pay-off functions, ϕ_1 and ϕ_2, and that we are looking for a so-called *equilibrium point* (x_0, z_0). Then (6.1.8) says that if player 1 (who must pay $\phi_1(x, z)$ if he selects x and his opponent selects z) deviates from x_0 while player 2 sticks to z_0, then the former will not have to pay less! And similarly for player 2.

This generalization is important, because we can now easily consider three or more players, whereas the saddle-point condition allows only two. That the latter indeed is a special case of the equilibrium point condition, is seen by taking $L = \phi_1 = -\phi_2$, so that $\phi_1 + \phi_2 = 0$, from which the name '*zero-sum game*' comes. (x_0, z_0)

being an equilibrium point can be shown to be equivalent to (x_0, z_0) being a fixed point of a suitably chosen mapping T, and it is easily seen that the condition $(x, z) \in C \times D$ can be replaced by the condition $(x, z) \in G$, where G is a more generally defined subset of $X \times Z$.

Returning to T, defined by (6.1.7), which is also termed *multi-function*, let us now try to answer the question: under what conditions does T have a fixed point? The question seems to be relevant as there are quite a few fixed point theorems available. All of them, however, share the unpleasant property that certain sets must be *compact*. On top of this, conditions of another nature must be imposed. By and large these are either *convexity* assumptions or that T must satisfy a certain *contraction* requirement. If T is actually a point-to-*point* mapping, a simple form of the latter is that

$$d(T(x, z),\ T(x', z')) \leqslant \rho d((x, z),\ (x', z')),$$

with d denoting the distance between two points, and with $0 \leqslant \rho < 1$. We shall first consider a contraction type fixed point theorem which is probably not of much use for our purposes, and then a fixed point theorem relying on convexity.

In what follows we take $S = C \times D$ and $s = (x, z)$. In the next theorem we shall assume that T is a point-to-point mapping, although a more general version is valid which involves the distance between points, and between points and sets, as well as the so-called *Hausdorff distance* between sets. We omit the proof because the use of the theorem seems limited.

6.1.9 **Theorem.** Let S be a complete metric space and let $T: S \to S$ satisfy the condition that for some nonnegative numbers a_i, $i = 1, \ldots, 5$, with $a_1 + a_2 + a_3 + a_4 + a_5 < 1$, and for all $s, s' \in S$,

$$d(T(s), T(s')) \leqslant a_1 d(s, T(s)) + a_2 d(s', T(s')) + a_3 d(s, T(s')) + a_4 d(s', T(s)) + a_5 d(s, s') \tag{6.1.10}$$

where d is the distance function. Then T has a fixed point.

The compactness condition we were talking about is contained in the requirement that $T(s)$ is a single point of S.

Consider now the following simple problem of finding

(6.1.11) $\inf_R(f(x)+g(x))$ with $f(x) = (x+1)^2$,
$g(x) = 3(x-1)^2$ and $\operatorname{dom} f = \operatorname{dom} g = [0, 1]$.

This is an extremely regular problem in that the domains of f and g are compact, and f and g are *strictly* convex, which means that

$$f(\lambda x+(1-\lambda)x') < \lambda f(x)+(1-\lambda)f(x')$$

if $x \neq x'$ and $0 < \lambda < 1$, and similarly for g. Approaching the problem from the standpoint of Fenchel duality, let the bifunction be $F(x, y) = f(x)+g(x+y)$. Then, if $0 \leqslant x \leqslant 1$,

(6.1.12)
$$\begin{aligned} L(x, z) &= (x+1)^2-zx-\tfrac{1}{12}z^2+z && \text{if } 0 \leqslant z \leqslant 6 \\ &= (x+1)^2-zx+z && \text{if } z \leqslant 0 \\ &= (x-1)^2-zx+3 && \text{if } z \geqslant 6. \end{aligned}$$

From (6.1.5) it follows that

(6.1.13*a*) $$B(x) = 6(1-x) \quad \text{for} \quad 0 \leqslant x \leqslant 1,$$

so that we can equally well restrict z to the interval $[0, 6]$, implying that $S = [0, 1] \times [0, 6]$ is compact. Furthermore we have that if $0 \leqslant z \leqslant 6$,

(6.1.13*b*)
$$\begin{aligned} A(z) &= \tfrac{1}{2}(z-2) && \text{if } 2 \leqslant z \leqslant 4 \\ &= 0 && \text{if } z \leqslant 2 \\ &= 1 && \text{if } z \geqslant 4. \end{aligned}$$

It follows that $T(0, 3) = (\frac{1}{2}, 6)$ and that $T(1, 3) = (\frac{1}{2}, 0)$. If we take $s = (0, 3)$ and $s' = (1, 3)$ the theorem requires that $6 \leqslant 3(a_1+a_2+a_3+a_4)+a_5$, which cannot be true. Yet we did not make a bad choice, for $z_0 = 3$ is the optimal value of z, although at the same time by taking $x = 0$ and $x = 1$ we avoided the optimal $x = x_0 = \frac{1}{2}$. Trying to replace T by T^n is of no help. For let $p = 2x-1$ and $q = \frac{1}{3}(z-3)$, so that $-1 \leqslant p, q \leqslant +1$. Then (p, q) is mapped on $(\bar{p}, \bar{q})$, with $\bar{q} = -p$, $\bar{p} = 3q$ if $-1 \leqslant 3q \leqslant 1$, $\bar{p} = -1$ if $3q \leqslant -1$ and $\bar{p} = 1$ if $3q \geqslant 1$. Applying T again and again we eventually arrive at the boundary of the p, q region, unless we take $(p, q) = (0, 0)$, corresponding to $(x, z) = (\frac{1}{2}, 3)$. Hence the technique of approximating the required fixed point, namely $(\frac{1}{2}, 3)$, by $T^n(x, z)$, where (x, z) is some starting point, is of no use.

We remark that other iterative schemes exist that converge nicely in this case. For example, given (x, z), compute $(\bar{x}, \bar{z})$ by $\bar{x} = x - \tau L_x(x, z)$ and $\bar{z} = z + \sigma L_z(x, z)$, where τ and σ are small positive parameters and L_x and L_z denote partial derivatives.

We now turn to a theorem based on convexity assumptions. It says that if S is convex and compact and if $T(s)$ is also convex for all $s \in S$, then there exists an $s_0 \in S$ such that $s_0 \in T(s_0)$, at least if in addition to these assumptions T satisfies a certain continuity requirement. There are several ways to demonstrate the truth of this statement. Below we use a combinatorial argument.

6.1.14 **Lemma** (Sperner). Let S be an n-dimensional simplex with vertices $p_0, p_1, \ldots, p_n$; that is let S be the set of all convex combinations $s = \sigma_0 p_0 + \sigma_1 p_1 + \ldots + \sigma_n p_n$ with $\sigma_j \geqslant 0$ and $\sigma_0 + \sigma_1 + \ldots + \sigma_n = 1$, and let $p_0 - p_1, \ldots, p_0 - p_n$ be independent, so that the σ_j are defined uniquely by s. Identify s with $(\sigma_0, \sigma_1, \ldots, \sigma_n)$, hence represent s by barycentric coordinates. Let $L: S \to \{0, 1, \ldots, n\}$ be a labelling of all elements s of S such that if $L(s) = j$ then $\sigma_j, > 0$. Let $S_1, S_2, \ldots$ be a finite number of (again n-dimensional) subsimplices of S such that $S = S_1 \cup S_2 \cup \ldots$ and such that if $i \neq k$, $S_i \cap S_k$ is either empty or a face (of any dimension) of both S_i and S_k. Denote by $L(S_i)$ the set of labels of all vertices of S_i. Then for some i we have that $L(S_i) = \{0, 1, \ldots, n\}$; that is the vertices of at least one subsimplex together carry all possible labels. We even claim that the number of such subsimplices is odd.

Proof. Observe that $L(p_j) = j$ for all j and that if $s_0, s_1, \ldots, s_{n-1}$ are n points all contained in one $(n-1)$-dimensional face of S and if $L(s_j) = j$, $j = 0, 1, \ldots, n-1$, then this face is the one spanned by $p_0, p_1, \ldots, p_{n-1}$. We shall use this observation below.

Call S_i *representing* if $L(S_i) = \{0, 1, \ldots, n\}$, and call an $(n-1)$-dimensional face of S_i *representing* if its vertices together carry the labels $0, 1, \ldots, n-1$. Let a be the number of representing subsimplices, let b_i be the number of representing faces of S_i, and let c be the total number of representing faces of $S_1, S_2, \ldots$ which are contained in an $(n-1)$-dimensional face of S.

If S_i is representing then $b_i = 1$. If it is not and $j \notin L(S_i)$ for some $j \leqslant n-1$, then $b_i = 0$. For all remaining cases $L(S_i) = \{0, 1, \ldots, n-1\}$ and one label occurs twice, so that then $b_i = 2$. It follows that $a \equiv \Sigma b_i$ (mod 2).

Next we claim that $c \equiv \Sigma b_i$ (mod 2). This is fairly obvious because each representing face that is inside S is the face of two subsimplices.

It then follows that $a \equiv c$ (mod 2). From the observation made at the beginning it follows that all representing faces are contained in the $(n-1)$-dimensional face of S that is spanned by $p_0, p_1, \ldots, p_{n-1}$. Hence if the theorem were true with n replaced by $n-1$, c would be odd and so a would be odd as well. By applying an inductive argument it remains only to be shown that the theorem is true for $n = 0$ and this is trivially the case.

6.1.15 **Lemma** (Brouwer). If S is an n-dimensional simplex and if $T: S \to S$ is a continuous (point-to-point) mapping, then $s_0 = T(s_0)$ for some $s_0 \in S$.

Proof. Introduce barycentric coordinates as in Lemma 6.1.14. If $s \in S$ let $s = (\sigma_0, \sigma_1, \ldots, \sigma_n)$ and let $s' = T(s) = (\sigma'_0, \sigma'_1, \ldots, \sigma'_n)$. Define the labelling $L: S \to \{0, 1, \ldots, n\}$ by $L(s) = \min\{j: \sigma'_j \leqslant \sigma_j, \sigma_j > 0\}$. $L(s)$ is properly defined since for each s there exists a j such that $\sigma_j > 0$ and such that $\sigma'_j \leqslant \sigma_j$, because the barycentric coordinates are nonnegative and sum to unity. Moreover L satisfies the requirement of Lemma 6.1.14. Consider now a sequence of subdivisions of S into subsimplices $S_{t1}, S_{t2}, \ldots$ such that the maximal 'diameter' of $S_{t1}, S_{t2}, \ldots$ tends to zero if t tends to infinity. Because S is compact and by Lemma 6.1.14 there is a subsequence of a sequence $\bar{S}_1 = S_{1i_1}, \bar{S}_2 = S_{2i_2}, \ldots$ such that $L(\bar{S}_t) = \{0, 1, \ldots, n\}$ for all t, and such that the vertices of $\bar{S}_t$ tend to a point s_0 for the subsequence. Now fix j and let s_t be the vertex of $\bar{S}_t$ with $L(s_t) = j$. Further let $s'_t = T(s_t)$, $s_t = (\sigma_{t0}, \ldots, \sigma_{tn})$, and $s'_t = (\sigma'_{t0}, \ldots, \sigma'_{tn})$. Then $\sigma'_{tj} \leqslant \sigma_{tj}$. But s_t converges to s_0, hence s'_t converges to $s'_0 = T(s_0)$, T being continuous. Hence, letting $s_0 = (\sigma_{o0}, \ldots, \sigma_{on})$ and $s'_0 = (\sigma'_{o0}, \ldots, \sigma'_{on})$, $\sigma'_{oj} \leqslant \sigma_{oj}$ for all j, so that $\sigma'_{oj} = \sigma_{oj}$ for all j, or $s_0 = T(s_0)$.

(The minimization in the definition of $L(s)$ only serves to define $L(s)$ uniquely.)

The next step is to extend the last lemma by taking for T a point-to-*set* mapping.

6.1.16 **Lemma** (Kakutani). If S is an n-dimensional simplex and if for all $s \in S$, $T(s)$ is a nonempty convex subset of S, then $s_0 \in T(s_0)$ for some $s_0 \in S$, if the set $G = \{(s, s'): s' \in T(s)\}$, which is the *graph* of T, is closed.

Proof. The condition on G can be reformulated as follows. If, as t tends to infinity, p_t tends to s_0 and $q_t \in T(p_t)$ tends to q_o, then $q_o \in T(s_0)$. The proof of this is straightforward and is left to the reader. We shall use this reformulation at the end of the proof.

Let $S_{t1}, S_{t2}, \ldots$ be as in the proof of Lemma 6.1.15. Define the point-to-*point* mapping T_t as follows. If v is a *vertex* of some S_{ti} let $T_t(v)$ be any element of $T(v)$, and if s is the convex combination $\lambda_0 v_0 + \lambda_1 v_1 + \ldots$ where v_0, v_1 are the vertices of some S_{ti}, let $T_t(s) = \lambda_0 T_t(v_0) + \lambda_1 T_t(v_1) + \ldots$ Then by Lemma 6.1.15 $s_t = T_t(s_t)$ for some $s_t \in S$. If we let t tend to infinity, a subsequence of the s_t tends to some point s_0. We want to show that $s_0 \in T(s_0)$.

Let S_{ti} be such that $s_t \in S_{ti}$ (hence i depends on t and a choice may be necessary here) and let $p_{to}, \ldots, p_{tn}$ be the vertices of S_{ti}, which all converge to s_0. Put $s_t = \lambda_{to} p_{to} + \ldots + \lambda_{tn} p_{tn}$ and $q_{tj} = T_t(p_{tj})$. Then $q_{tj} \in T(p_{tj})$ and $s_t = T_t(s_t) = \lambda_{to} q_{to} + \ldots + \lambda_{tn} q_{tn}$. Let q_{tj} converge to q_{oj} and λ_{tj} to λ_{oj} (for some subsequence). Since p_{tj} tends to s_0 we have that s_t tends to

$$\lambda_{oo} s_0 + \ldots + \lambda_{on} s_0 = s_0 = \lambda_{oo} q_{oo} + \ldots + \lambda_{on} q_{on}.$$

And since p_{oj} tends to s_0 and $q_{tj} \in T(p_{tj})$ tends to q_{oj} it follows from the reformulation stated at the beginning that $q_{oj} \in T(s_0)$ for all j. But $s_0 = \lambda_{oo} q_{oo} + \ldots + \lambda_{on} q_{on}$ and $T(s_0)$ is convex; hence $s_0 \in T(s_0)$.

We remark that the graph of T is closed if and only if $T(s)$ is compact for all $s \in S$, and for all $s \in S$ and all open sets $V \supset T(s)$ there exists a neighbourhood $U(s)$ of s such that $T(s') \subset V$ if $s' \in U(s)$. This gives a third formulation of the condition we imposed on T. In this form (but also in the other forms) it is sometimes called *upper semi-continuity* of T (a term we have been using before for something different). Then T is called *lower semi-continuous* if for all $s \in S$ and all open sets V

with $V \cap T(s) \neq \varnothing$ there exists a neighbourhood $U(s)$ of s such that $V \cap T(s') \neq \varnothing$ if $s' \in U(s)$. Compactness of $T(s)$ is not required in the latter definition. Both definitions reduce to ordinary continuity if T is a point-to-point mapping (showing the difference of these definitions with what we called u.s.c. and l.s.c. before).

6.1.17 **Corollary.** Lemma 6.1.16 remains true if S is replaced by any nonempty compact convex subset of R_n.

Proof. There exists a simplex $S' \subset R_n$ such that $S' \supset S$. Let s_1 be some element of ri S, and for all directions $s_2 \neq 0$ let $\lambda_2 = \max\{\lambda: s_1 + \lambda s_2 \in S\}$. Define $f: S' \to S$ by $f(s') = s'$ if $s' \in S$, and $f(s') = f(s_1 + \lambda_2 s_2)$ if $s' \notin S$ and $s' = s_1 + \lambda s_2$ for some $s_2 \neq 0$ and some $\lambda > 0$. This (*retracting*) mapping is continuous and it is not difficult to show that the graph of Tf is closed, hence $s'_0 \in T(f(s'_0))$ for some $s'_0 \in S'$. But $T(f(s'_0)) \subset S$ so that $s'_0 \in S$ and hence $f(s'_0) \in S$, from which we see that $s'_0 \in T(s'_0)$.
At last we can now jump to the desired result.

6.1.18 **Theorem** (Glicksberg, Tychonoff, Kakutani, Ky Fan). Let X be a separated (or Hausdorff) locally convex topological vector space. Let S be a nonempty compact convex subset of X and let, for all $s \in S$, $T(s)$ be a nonempty convex subset of S. Then $s_0 \in T(s_0)$ for some $s_0 \in S$ if the graph of T, which is the set $G = \{(s, s'): s' \in T(s)\}$, is closed.

Proof. Let W be any neighbourhood of the origin, such that W is convex and closed, and such that $W = -W$.

Since S is compact there exist $p_0, \ldots, p_n \in S$ such that $S \subset \bigcup_j (p_j + W)$. Let S' be the set of all convex combinations of these points. Then $S' \subset S$ since S is convex. Define the mapping T'_W from S' to its subsets by $T'_W(s) = (T(s) + W) \cap S'$, $s \in S'$. Then we have

(*a*) $T'_W(s) \neq \varnothing$; for $T(s) \neq \varnothing$, and $s' \in T(s)$ implies that $s' = p_j - w$ for some j and some $w \in -W = W$, so that $s' + w = p_j \in S'$ and $s' + w \in T'_W(s)$.

(*b*) $T'_W(s)$ is convex; for $T(s)$, W and S' are all convex.

(*c*) The graph of T'_W is closed. The proof of this uses the reformulation mentioned at the beginning of the proof of Lemma 6.1.16. Let s_t tend to $s_0 \in S'$ and let $s'_t \in T'_W(s_t)$ tend to $s'_0 \in S'$, as

t tends to infinity. Then $s'_t = s''_t + w_t$ for some $s''_t \in T(s_t)$ and some $w_t \in W$. But $T(s_t) \subset S$, which is compact, hence s''_t tends to some s''_0 (for a subsequence), so that w_t tends to $s'_0 - s''_0 \in W$, because W is closed. Since the graph of T is closed by assumption, it follows that $s''_0 \in T(s_0)$, so that $s'_0 = s''_0 + (s'_0 - s''_0) \in T(s_0) + W$ and since $s'_0 \in S'$ we have that $s'_0 \in T'_W(s_0)$.

By Corollary 6.1.17, $s_0 \in T'_W(s_0)$ for some $s_0 \in S' \subset S$. In other words the set $Q(W) = \{s: s \in (T(s) + W) \cap S\}$ is not empty. Moreover $Q(W)$ is closed. The proof of the latter is quite similar to that of showing that the graph of T'_W is closed (see (*c*) above).

Finally let $W = W_t$ such that $W_{t+1} \subset W_t$ and such that W_t tends to $\{0\}$ as t tends to infinity. Then $Q(W_{t+1}) \subset Q(W_t)$ and since $Q(W_t)$ is closed for all t and is contained in the compact set S, it follows that $s_0 \in Q(W_t)$ for some $s_0 \in S$ and all t. From this it easily follows that s_0 must be an element of $T(s_0)$.

6.1.19 **Example.** Consider again the (strictly convex) problem of finding the infimum of (6.1.11), where $S = [0, 1] \times [0, 6]$ and T is a point-to-point mapping defined by

$$\begin{aligned} T(x, z) &= (\tfrac{1}{2}(z-2), 6(1-x)) && \text{if } 2 \leqslant z \leqslant 4 \\ &= (0, 6(1-x)) && \text{if } z \leqslant 2 \\ &= (1, 6(1-x)) && \text{if } z \geqslant 4. \end{aligned}$$

It is not difficult to verify the conditions of the theorem, hence, contrary to a previous attempt, we may conclude that T has a fixed point and that L has a saddle-point. In this simple case it is very easy to find that if $(x_o, z_0) = T(x_o, z_0)$ then $x_o = \frac{1}{2}$ and $z_0 = 3$.

Applying the theorem to linear programming is not so successful. Let $\alpha = \inf\{cx: Ax \geqslant b, x \geqslant 0\}$ and $L(x, z) = cx + z(b - Ax)$ if $x \geqslant 0$ and $z \geqslant 0$. Then if the set $A(z)$ is to be nonempty we must require that $zA \leqslant c$ (and $z \geqslant 0$), and then $A(z) = \{x: (c - zA)x = 0,\ x \geqslant 0\}$. Similarly $B(x) \neq \varnothing$ only if $Ax \geqslant b$ (and $x \geqslant 0$) and then $B(x) = \{z: z(b - Ax) = 0, z \geqslant 0\}$. The following problems now arise. First of all $A(z)$ and $B(x)$ may well be unbounded for certain z and x. Moreover if we restrict x to vectors satisfying $Ax \geqslant b$ and $x \geqslant 0$, then it is questionable whether $A(z)$ is still nonempty for z satisfying $zA \leqslant c$ and $z \geqslant 0$, and something similar may be said if the roles of x and z are interchanged. But even if everything is fine in these respects then we are still faced with the problem that at least one of the feasible regions $G = \{x: Ax \geqslant b,\ x \geqslant 0\}$ and $G^* = \{z: zA \leqslant c,\ z \geqslant 0\}$ is unbounded, or more precisely: if for example G is non-empty and bounded then G^* is nonempty but unbounded. To see this choose $e < 0$ such that $c+e < 0$. From the duality theorem of linear programming it follows that there exists a $z \geqslant 0$ such that $zA \leqslant c+e < 0$, so that $z \neq 0$, and if $\omega \geqslant 1$, then $\omega zA \leqslant c+e+(\omega-1)(c+e) < c$, so that $\omega zA \leqslant c$ for all $\omega \geqslant 1$ and for some $z \geqslant 0$, $z \neq 0$, implying that G^* is unbounded.

It would seem, therefore, that something like *strict* convexity is required if we want to be able to apply our fixed point theorem.

When we are interested in establishing the existence of a saddle-point of a given Lagrangian function and not so much in solving a pair of optimization problems, then the theorem opens the way to conditions weaker than convexity and concavity. So consider spaces X and Z, convex sets $C \subset X$ and $D \subset Z$ and a function $L{:}S = C \times D \to R$. Taking the conditions of the theorem about sets being compact or closed for granted, we must specify conditions to ensure that $T(x, z)$ is a convex set. The following will do:

(6.1.20) The sets $\{x': L(x', z) \leqslant \gamma,\ x' \in C\}$ and

$\{z': L(x, z') \geqslant \delta,\ z' \in D\}$ are convex for all $z \in D$,

all $x \in C$ and all $\gamma, \delta \in R$.

This leads to the next definition.

6.1.21 **Definition.** We say that a function $h: Z \to R$ is *quasi-convex* if the sets $\{z': h(z') \leqslant \gamma\}$, which are called the *level sets of* h, are convex for all real γ. The function h is *quasi-concave* if $-h$ is quasi-convex. Furthermore we say that $L(x, z)$ is *quasi-convex-concave* (on $C \times D$) if $L(\cdot, z)$ is quasi-convex (on C) for all $z \in D$, and if $L(x, \cdot)$ is quasi-concave (on D) for all $x \in C$.

It easily follows from this definition that if h is quasi-convex then the set $A = \{z': h(z') = \min_Z h(z)\}$ is a convex set. For take $\gamma = \min_Z h(z)$, let z_1' and z_2' be elements of A and $0 \leqslant \lambda \leqslant 1$. Then $h(\lambda z_1' + (1-\lambda)z_2') \leqslant \gamma = \min_Z h(z)$, and here we may, of course, replace the inequality by equality. From this it now follows that the conditions (6.1.20) are indeed sufficient to ensure that $T(x, z)$ is convex, because (6.1.20) says nothing other than that L is quasi-convex-concave, and $T(x, z) = A(z) \times B(x)$ with $A(z)$ and $B(x)$ defined as in (6.1.5).

Since quite a few nonconvex functions, even certain concave functions, are quasi-convex we can expect that through the use of fixed point theorems we can claim the existence of saddle-points for a much wider class of Lagrangians than before. As we said earlier this only applies to situations where we are interested in Lagrangians as such. As soon as we are interested in solving a given primal problem and a corresponding dual problem, then $L(x, z)$ is automatically concave in z if we require that the Lagrangian can be obtained by conjugation from a given bifunction, and $L(x, z)$ is automatically convex in x if we require that the Lagrangian can be obtained by conjugation from the dual bifunction.

The foregoing discussion immediately leads to a *minimax theorem.*

6.1.22 **Theorem.** Let C and D be nonempty compact convex subsets of separated locally convex topological vector spaces X and Z, respectively. Let $L: C \times D \to R$ be quasi-convex-concave and be continuous in the pair (x, z). Then

(6.1.23) $$\min_C \max_D L(x, z) = \max_D \min_C L(x, z).$$

Proof. The proof is straightforward, except perhaps for two points. Let $S = C \times D$, $A(z)$ and $B(x)$ as in (6.1.5), and $T(x, z) = A(z) \times B(x)$.

(A) In order to show that $T(x, z) \neq \varnothing$ we must show that $A(z) \neq \varnothing$ and $B(x) \neq \varnothing$ if $z \in D$ and $x \in C$, and this follows from the compactness and continuity assumptions.

(B) And in order to show that the graph of T is closed we must show that if

$$L(x_i', z_i) = \min_C L(x, z_i), \quad L(x_i, z_i') = \max_D L(x_i, z),$$

and
$$(x_i, z_i, x_i', z_i') \to (x_o, z_o, x_o', z_o')$$
then
$$L(x_o', z_o) = \min_C L(x, z_o) \quad \text{and} \quad L(x_o, z_o') = \max_D L(x_o, z),$$

and this too follows from the compactness and continuity assumptions.

Notice, however, that in order to show the claim made under (A) it is sufficient that $L(\cdot, z)$ is l.s.c for all $z \in D$ and that $L(z, \cdot)$ is u.s.c for all $x \in C$. Further notice that a completely different proof of Theorem 6.1.22 will be given in the next section; see Corollary 6.2.10.

6.2 The sup inf = inf sup approach

If we are willing to give up the existence of a saddle-point and are content with that of a saddle-value, then we can relax the compactness conditions we have been using in 6.1.

As before we shall assume that X and Z are separated locally convex topological vector spaces, and that C and D are convex subsets of X and Z, respectively, with D compact but C not necessarily so (symmetric results are obtained by taking C compact).

6.2.1 **Definition.** We say that $L: C \times D \to R$ is *l.s.c.-u.s.c* if $L(\cdot, z)$ is lower semi-continuous for all $z \in D$, and if $L(x, \cdot)$ is upper semi-continuous for all $x \in C$.

The theorem we want to prove is the following.

6.2.2 **Theorem.** Let C and D be nonempty convex subsets of separated locally convex topological vector spaces X and Z, respectively, with D compact. Let $L: C \times D \to R$ be quasi-convex-concave and l.s.c.-u.c.s. Then

$$\sup_D \inf_C L(x, z) = \inf_C \sup_D L(x, z).$$

The proof of this theorem is based among other things on the finite intersection property of a family of closed subsets of a compact set D.

6.2.3 **Definition.** A family of sets $M(x)$ with $x \in C$ is said to have the *finite intersection property* if

$$M(x_1) \cap \ldots \cap M(x_k) \neq \varnothing$$

for any finite subset $\{x_1, \ldots, x_k\} \subset C$.

6.2.4 **Lemma.** If, for all $x \in C$, $M(x)$ is a closed subset of the compact set D and if this family of sets has the finite intersection property then $z \in M(x)$ for some $z \in D$ and all $x \in C$.

Proof. If the required z did not exist the complements of $M(x)$, which are open, would cover D; hence a finite number of them would do so (since D is compact) thus implying that a finite number of the given family would not intersect.

Define α by $\alpha = \inf_C \sup_D L(x, z)$. Then if $\alpha = -\infty$ the theorem trivially holds. But we may also forget about the case where $\alpha = +\infty$, because of the next result.

6.2.5 **Lemma.** If $L(x, \cdot)$ is u.s.c. and if D is compact then $L(x, \cdot)$ is bounded above.

Proof. If $L(x, \cdot)$ were not bounded above then $L(x, z_i)$ would tend to plus infinity for a sequence z_i tending to z, but because of the upper semi-continuity $L(x, z_i) \leqslant L(x, z)+1$ for i large enough, implying that $L(x, z) = +\infty$.

Next let us introduce the following level sets.

6.2.6 **Definition.** $M(x, \gamma) = \{z : z \in D,\ L(x, z) \geqslant \gamma\}$.

6.2.7 **Lemma.** For all $\epsilon > 0$, $M(x, \alpha - \epsilon)$ is nonempty and compact if D is compact and if $L(x, \cdot)$ is u.s.c.

Proof. The definition of α immediately implies that $M(x, \alpha - \epsilon) \neq \varnothing$. The compactness of $M(x, \alpha - \epsilon)$ follows if we can show that this set is closed, for it is a subset of the compact set D. So let $z_i \in D$ tend to z and let $L(x, z_i) \geqslant \alpha - \epsilon$. By the upper semi-continuity $L(x, z_i) \leqslant L(x, z) + \epsilon'$ for all $\epsilon' > 0$ and i large enough, so that $L(x, z) \geqslant \alpha - \epsilon - \epsilon'$, hence that $L(x, z) \geqslant \alpha - \epsilon$.

The next lemma is the first serious step towards the proof of the theorem.

6.2.8 **Lemma.** Let X, Z, C and D be as in the theorem, and let $L(x, \cdot)$ be u.s.c. for all $x \in C$. Then

$$\sup_D \inf_C L(x, z) = \inf_C \sup_D L(x, z)$$

if there exists a sequence $\epsilon_n > 0$ such that ϵ_n tends to zero and such that for each n the family of sets $M(x, \alpha-\epsilon_n)$ with $x \in C$ has the finite intersection property.

Proof. By Lemmas 6.2.4 and 6.2.7, for all n there exists a $z_n \in D$ such that $z_n \in M(x, \alpha-\epsilon_n)$ for all $x \in C$, implying that

$$\inf_C L(x, z_n) \geqslant \alpha - \epsilon_n \quad \text{and that} \quad \sup_D \inf_C L(x, z) \geqslant \alpha.$$

Since $\sup_D \inf_C L(x, z) \leqslant \alpha$ always the proof is complete.

Notice that up to now we have not used the quasi-convex-concavity of L. But in the next and last lemma we shall use it to show that a sequence like the one in Lemma 6.2.8 exists, and this at the same time will prove the theorem.

6.2.9 **Lemma.** If the conditions of Theorem 6.2.2 are true then for any $\epsilon > 0$ the family of sets $M(x, \alpha-\epsilon)$ with $x \in C$ has the finite intersection property.

Proof. The proof is by induction with respect to the number of sets $M(x_i, \alpha-\epsilon)$, $x_i \in C$, $i = l, \ldots, k$.

(A) First we take $k = 2$ and put $x(\lambda) = (1-\lambda)x_1 + \lambda x_2$ for $0 \leqslant \lambda \leqslant 1$, and $M(\lambda) = M(x(\lambda), \alpha-\epsilon)$.

Contrary to what has to be shown suppose that $M(0) \cap M(1) = \emptyset$. Then $M(\lambda) \subset M(0) \cup M(1)$ for all λ, $0 \leqslant \lambda \leqslant 1$; for since $L(\cdot, z)$ is quasi-convex it is not possible that

$$\begin{aligned} L(x(0), z) &< \alpha - \epsilon, \\ L(x(\lambda), z) &\geqslant \alpha - \epsilon, \\ L(x(1), z) &< \alpha - \epsilon, \end{aligned}$$

so that if $L(x(\lambda), z) \geqslant \alpha - \epsilon$ for some z, $L(x(0), z) \geqslant \alpha - \epsilon$ or $L(x(1), z) \geqslant \alpha - \epsilon$.

We can now go one step further and claim that $M(\lambda) \subset M(0)$ or $M(\lambda) \subset M(1)$, for otherwise for some z_1 and z_2 we would have $z_1 \in M(\lambda) \cap M(0)$ and $z_2 \in M(\lambda) \cap M(1)$, so that

$$\begin{aligned} L(x(\lambda), z_1) &\geqslant \alpha - \epsilon, \\ L(x(\lambda), z_2) &\geqslant \alpha - \epsilon. \end{aligned}$$

Hence, by the quasi-concavity of $L(x, \cdot)$,

$$L(x(\lambda),(1-\mu)z_1+\mu z_2) \geqslant \alpha-\epsilon \quad \text{for all } \mu, \quad 0 \leqslant \mu \leqslant 1,$$

and so $$(1/\mu)z_1+\mu z_2 \in M(\lambda) \subset M(0) \cup M(1).$$

But $M(0)$ and $M(1)$ are closed (by Lemma 6.2.7) and convex, so that for some μ_o,
$$(1-\mu_o)z_1+\mu_o z_2 \in M(0) \cap M(1),$$

contradicting $M(0) \cap M(1) = \varnothing$.

Moreover if $M(\lambda) \subset M(0)$ and if $0 < \mu < \lambda$ then $M(\mu) \subset M(0)$, for otherwise $M(\mu) \subset M(1)$ so that for some $z \in M(1)$,

$$L(x(0), z) < \alpha-\epsilon,$$
$$L(x(\mu), z) \geqslant \alpha-\epsilon,$$
$$L(x(\lambda), z) < \alpha-\epsilon,$$

contradicting the quasi-convexity of $L(\cdot, z)$.

We conclude therefore that for some μ_o $M(\mu) \subset M(0)$ if $\mu \leqslant \mu_o$ or if $\mu < \mu_o$, and $M(\mu) \subset M(1)$ if $\mu > \mu_o$ or if $\mu \geqslant \mu_o$, respectively. We need consider only one of these two possibilities. Let us take the first one. In the case $z \notin M(\mu_o)$ then $L(x(\mu_o), z) < \alpha-\epsilon$ as follows from the definition of $M(\lambda)$. And when $z \in M(\mu_o)$ then $L(x(\mu_o), z) \leqslant \alpha-\epsilon$. To see the latter take $\mu > \mu_o$ and let μ tend to μ_o. Then, by the lower semi-continuity of $L(\cdot, z)$, for all $\epsilon' > 0$, $L(x(\mu), z) \geqslant L(x(\mu_o), z)-\epsilon'$ if μ is sufficiently close to μ_o. But $L(x(\mu), z) < \alpha-\epsilon$ so that $L(x(\mu_o), z) < \alpha-\epsilon+\epsilon'$ and $L(x(\mu_o), z) \leqslant \alpha-\epsilon$.

The final conclusion now is that $L(x(\mu_o), z) \leqslant \alpha-\epsilon$ for all $z \in D$ and some μ_o. But this contradicts the definition of α, for it means that $\inf_C \sup_D L(x, z) \leqslant \alpha-\epsilon$. Hence we are unable to avoid all the contradictions encountered in this part of the proof and so must conclude that $M(0) \cap M(1) \neq \varnothing$.

(B) Assume that the finite intersection property holds for $k-1$ sets $M(x_i, \alpha-\epsilon)$ with $k-1 \geqslant 2$. Let us now verify this property for k such sets. Define

$$M'(x, \alpha-\epsilon) = M(x, \alpha-\epsilon) \cap M(x_k, \alpha-\epsilon), \quad x \in C$$

and $D' = M(x_k, \alpha-\epsilon)$. Consider the problem of proving that $\alpha' = \inf_C \sup_{D'} L(x, z)$ equals $\sup_{D'} \inf_C L(x, z)$. Clearly $\alpha' \leqslant \alpha$, and in fact $\alpha' = \alpha$. For let $\epsilon_n > 0$ and let ϵ_n tend to zero, then by (A)

$M'(x, \alpha-\epsilon_n) \neq \emptyset$ for all $x \in C$, which means that $\alpha' \geqslant \alpha-\epsilon_n$ so that $\alpha' \geqslant \alpha$. Since all conditions of the theorem remain true for the reduced problem (that is with D replaced by D') it follows from the induction assumption that

$$M'(x_1, \alpha-\epsilon) \cap \ldots \cap M'(x_{k-1}, \alpha-\epsilon) \neq \emptyset,$$

from which the statement of the lemma immediately follows.

6.2.10 **Corollary** (a reformulation of Theorem 6.1.22). If the conditions of Theorem 6.2.2 hold, if moreover L is continuous in the pair of variables (x, z) and if C, too, is compact, then L has a saddle-point.

Proof. Let γ be the saddle-value which exists by Theorem 6.2.2 and which is finite by Lemma 6.2.5. Then because of the definition of γ, the continuity of L (l.s.c.-u.s.c. is sufficient here) and the compactness of C and D there exist sequences

$$x_i \to x_o, \quad z'_i = z(x_i) \to z'_0, \quad z_i \to z_0, \quad x'_i = x(z_i) \to x'_o$$

such that both $L(x_i, z'_i)$ and $L(x'_i, z_i)$ tend to γ and such that

$$L(x_i, z'_i) = \max_D L(x_i, z) \quad \text{and} \quad L(x'_i, z_i) = \min_C L(x, z_i).$$

By the continuity of L it follows from this that

$$L(x_o, z'_0) = \max_D L(x_o, z) \quad \text{and that} \quad L(x'_o, z_0) = \min_C L(x, z_0),$$

so that $\gamma = L(x_o, z'_0) = \max_D L(x_o, z) = L(x'_o, z_0) = \min_C L(x, z_0)$; and hence (x_o, z_0) is a saddle-point of L.

6.3 Modified Lagrangians

Recall that given the problem of finding

$$\alpha = \inf\{f(x): g(x) \leqslant 0, \quad h(x) = 0\}$$

the Lagrangian function is

$$L(x, y^*, z^*) = f(x) + y^*g(x) + z^*h(x) \quad \text{for} \quad y^* \geqslant 0$$

and that we wanted to know under what conditions α is equal to

$$\beta = \sup_{y^* \geqslant 0, z^*} \inf_x L(x, y^*, z^*).$$

Suppose that for some x_o, y_o^*, z_o^* we have that

(6.3.1)
$$L'(x_o, y_o^*, z_o^*) = 0, \quad y_o^* \geqslant 0, \quad g(x_o) \leqslant 0, \quad h(x_o) = 0, \quad y_o^* g(x_o) = 0$$

and that

(6.3.2)
$$L''(x_o, y_o^*, z_o^*)$$

is positive definite on the subspace

$$S = \{x: g_a'(x_o)x = 0, h'(x_o)x = 0\},$$

where primes indicate differentiation with respect to x, and g_a is the part of g that is active at x_o. These conditions do not, of course, ensure that x_o is a local minimum of $L(x, y_o^*, z_o^*)$ for the simple reason that in (6.3.2) x is restricted to a certain subspace. Can we modify L in such a way that (6.3.1) remains true and that the positive definiteness of $L''(x_o, y_o^*, z_o^*)$ holds for the entire space X? If so then x would be a local minimum of the *modified Lagrangian* at the point (x_o, y_o^*, z_o^*) and we could take advantage of this fact when constructing algorithms for solving for x_o (assuming for the moment that y_o^* and z_o^* are known).

Instead of the term 'modified Lagrangian' several authors speak of '*augmented Lagrangian*' perhaps to suggest that certain terms are added to L (which in a sense is true) and sometimes the term 'generalized Lagrangian' is used.

One of the simpler proposals out of many is to replace L by the modified form

(6.3.3)
$$L_m(x, y^*, z^*, v, w) = f(x) + \Sigma_i y_i^* [(1+g_i(x))^{1+v_i} - 1]/(1+v_i) + \Sigma_j z_j^* h_j(x) + \Sigma_j w_j h_j^2(x)$$

with $v = (v_1, v_2, \ldots)$, $w = (w_1, w_2, \ldots)$, $v_i, w_j \in R$, and v_i even. Here we assume that g, as well as h, consists of finitely many scalar constraints. In fact we already made this assumption when we wrote down (6.3.2), at least as far as g is concerned, since we were talking about the active part of g at x_o. The computation of L_m' and L_m'' is straightforward and it is not difficult to verify that if x_o, y_o^*, z_o^* satisfy (6.3.1), then

(6.3.4) $$L'_m(x_o, y_o^*, z_o^*, v, w) = 0$$

and

(6.3.5) $$L''_m(x_o, y_o^*, z_o^*, v, w)xx = L''(x_o, y_o^*, z_o^*)xx + \Sigma_i y_{oi}^* v_i [g_i'(x_o)x]^2 + 2\Sigma_j w_j [h_j'(x_o)x]^2,$$

where we take i such that $g_i(x_o) = 0$. Now suppose that (6.3.2) is satisfied as well. Then it follows immediately that $L''_m(x_o, y_o^*, z_o^*, v, w)$ is positive definite on the subspace S, so if $x \notin S$ then for some i $g_i(x_o) = 0$ and $g_i'(x_o)x \neq 0$ or for some j $h_j'(x_o)x \neq 0$ (or both). But the last two terms on the right of (6.3.5) contain squares so if we further assume that

(6.3.6) $$y_{oi}^* > 0 \quad \text{if} \quad g_i(x_o) = 0$$

and if we take v and w *positive* and *large enough* then we have that $L''_m(x_o, y_o^*, z_o^*, v, w)$ is positive definite on the entire space X, at least if the set $\{x: |x| \leq 1\}$ is compact, and the latter is the case if X is reflexive. The proof of this statement is best given by an indirect argument and by restricting x to the unit ball.

Hence we may conclude that x is a local minimum of $L_m(x, y_o^*, z_o^*, v, w)$ if x_o, y_o^*, z_o^* satisfy the first-order and second-order conditions (6.3.1) and (6.3.2) as well as condition (6.3.6), if the number of constraints is finite, if X is reflexive, if v and w are positive and large enough, and, of course, if everything is twice differentiable.

Furthermore, if in addition to all these assumptions we assume that the second-order derivatives are continuous in (x, y^*, z^*) then L_m is strictly convex in x for (x, y^*, z^*) in a neighbourhood of (x_o, y_o^*, z_o^*).

Conversely, assume that the first-order conditions hold for L and hence for L_m, and that $L''_m(x_o, y_o^*, z_o^*, v, w)$ is positive definite for some $(v, w) > 0$. Then obviously $L''(x_o, y_o^*, z_o^*)$ is positive definite on the subspace S. And by Theorem 5.3.4, extended to include inequality as well as equality constraints, it follows that x_o is a strict (or '*isolated*') local minimum of the initial problem.

The conclusion, therefore, is that under certain conditions solving the primal problem for x_o amounts to finding (x_o, y_o^*, z_o^*) such that $y_o^* \geq 0$, $g(x_o) \leq 0$, $h(x_o) = 0$, $y_o^* g(x_o) = 0$ and

$$L_m(x_o, y_o^*, z_o^*, v, w) = \min_X L_m(x, y_o^*, z_o^*, v, w)$$

for suitable $(v, w) > 0$. Naturally one of the problems would be how large to choose v and w.

The discussion on the function L_m defined by (6.3.3) gives us enough hints to generalize. Deliberately, but reasonably, let

(6.3.7)
$$L_m(x, y^*, z^*, v, w) = f(x) + \Sigma_i p_i(g_i(x), y_i^*, v_i) + \Sigma_j q_j(h_j(x), z_j^*, w_j)$$

with x in a Banach space X and $y^* = (y_1^*, y_2^*, \ldots)$, $v = (v_1, v_2, \ldots)$, $z^* = (z_1^*, z_2^*, \ldots)$, $w = (w_1, w_2, \ldots)$ in finite-dimensional spaces, and stipulate that

(1) $p_i'(a, y_i^*, v_i) = y_i^*$ for all $v_i \geqslant 0$ and all $a \leqslant 0$, $y_i^* \geqslant 0$ such that $ay_i^* = 0$,
(2) $q_j'(0, z_j^*, w_j) = z_j^*$ for all $w_j \geqslant 0$,
(3) $p_i''(a, 0, v_i) = 0$ for all $v_i \geqslant 0$ and all $a < 0$,
(4) $p_i''(0, y_i^*, v_i) \to +\infty$ if $v_i \to +\infty$ for all $y_i^* \geqslant 0$,
(5) $q_j''(0, z_j^*, w_j) \to +\infty$ if $w_j \to +\infty$ for all z_j^*.

Differentiation in (1) to (5) is with respect to the first argument of p_i or q_j, *not* with respect to x.

Notice that the example of (6.3.3) does not satisfy condition (4) since in this condition we allow $y_i^* = 0$. We shall come back to this later.

It follows that

(6.3.8)
$$L_m'(x_o, y_o^*, z_o^*, v, w) = f'(x_o) + \Sigma_i p_i'(g_i(x_o), y_{oi}^*, v_i) g_i'(x_o) + \Sigma_j q_j'(h_j(x_o), z_{oj}^*, w_j) h_j'(x_o),$$

and from this and (1) and (2) that if x_o, y_o^*, z_o^* satisfy the conditions (6.3.1) and if $(v, w) \geqslant 0$

(6.3.9) $$L_m'(x_o, y_o^*, z_o^*, v, w) = L'(x_o, y_o^*, z_o^*) = 0.$$

Further, we have, still assuming that (6.3.1) holds, and that $(v, w) \geqslant 0$, that

(6.3.10) $$L_m''(x_o, y_o^*, z_o^*, v, w)xx = L''(x_o, y_o^*, z_o)xx + \Sigma_i p_i''(g_i(x_o), y_{oi}^*, v_i)[g_i'(x_o)x]^2 + \Sigma_j q_j''(h_j(x_o), z_{oj}^*, w_j)[h_j'(x_o)x]^2.$$

If $L''_m(x_o, y^*_o, z^*_o, v, w)$ is positive definite and if x is in the subspace S defined in (6.3.2) then $L''(x_o, y^*_o, z^*_o)xx > 0$ if $x \neq 0$, as follows from (6.3.10) and (3). Conversely, if x_o, y^*_o, z^*_o satisfy conditions (6.3.1) and (6.3.2) then $L''_m(x_o, y^*_o, z^*_o, v, w)$ is positive definite on S and even on the entire space X if X is reflexive and if the components of v and w are taken positive and large enough. For suppose the latter were not true. Then for $k = 1, 2, \ldots$ there would be $x^k \notin S$ with $|x^k| = 1$ and converging to some x' with $|x'| = 1$, and v^k and w^k with v^k_i and w^k_j tending to $+\infty$ for all i and all j, such that the right-hand side of (6.3.10) with $x = x^k$, $v_i = v^k_i$, $w_j = w^k_j$ is not positive. Because of condition (3) and the fact that $y^*_o g(x_o) = 0$, only terms with $g_i(x_o) = 0$ remain in the sum over i. And since x^k tends to x' and because of (4) and (5) the said right-hand side will only remain finite if $g'_i(x_o)x^k$ tends to zero for any i such that $g_i(x_o) = 0$, and if $h'_j(x_o)x^k$ tends to zero for all j. This would imply that $x' \in S$, so that the right-hand side would converge to something positive: a contradiction.

We have already remarked that condition (4) precludes the possibility of defining L_m as in (6.3.3). If we want to include this possibility and others then we must in (4) replace the condition that $y^*_i \geq 0$ by $y^*_i > 0$ and at the same time require that condition (6.3.6) holds, so that we have '*strict complementarity*'. This is easily verified.

From what we have said the following theorem is obtained.

6.3.11 **Theorem.** Let x_o be an element of a reflexive Banach space X, and let y^*_o, z^*_o, v and w be elements of finite-dimensional spaces such that y^*_o and v, as well as z^*_o and w have the same dimension, and such that x, y^*_o, z^*_o satisfy the first-order conditions (6.3.1). Further, let L_m be defined as in (6.3.7) satisfying the conditions (1) to (5), where (4) may be relaxed by letting $y^*_i > 0$ if $g_i(x_o) = 0$ implies that $y^*_{oi} > 0$ (condition (6.3.6)). Finally let everything be twice Fréchet-differentiable.

Then $L'_m(x_o, y^*_o, z^*_o, v, w) = 0$ for all v and all w and $L''_m(x_o, y^*_o, z^*_o, v, w)$ is positive definite for positive and sufficiently large v and w, if and only if $L''(x_o, y^*_o, z_o)$ is positive definite on the subspace S defined in (6.3.2).

It would be interesting to view the idea of introducing modifications from the standpoint of bifunctions. Can we construct a modified bifunction $F_m(x, y, z, v, w)$ such that

$$(6.3.12)\quad L_m(x, y^*, z^*, v, w) = \inf_{Y,Z}(F_m(x, y, z, v, w) + y^*y + z^*z)?$$

In a number of cases this is indeed possible. For example, in order to arrive at (6.3.3) we can put

$$(6.3.13)\quad \begin{aligned} F_m(x, y, z, v, w) &= f(x) + \Sigma_j w_j h_j^2(x) \\ &\qquad \text{if} \quad ((1+g_i(x))^{1+v_i} - 1)/(1+v_i) \leqslant y_i \\ &\qquad \text{and} \quad h(x) = z, \\ &= +\infty \text{ otherwise.} \end{aligned}$$

On the other hand there exist cases where it is not possible to apply (6.3.12), for example when L_m is strictly convex in z^* for $w > 0$, because (6.3.12) always yields an L_m which is concave in z^*.

Formula (6.3.13) suggests we might consider the following modified bifunction (assuming that only inequality constraints are present)

$$(6.3.14)\quad \begin{aligned} F_m(x, y, v) &= f(x) + \Sigma_i v_i y_i^2 \quad &\text{if} \quad g_i(x) \leqslant y_i, \\ &= +\infty &\text{otherwise.} \end{aligned}$$

Applying (6.3.12) we obtain

$$(6.3.15)\quad L_m(x, y^*, v) = f(x) + \Sigma_i p_i(g_i(x), y_i^*, v_i)$$

with
$$\begin{aligned} p_i(a, y_i^*, v_i) &= y_i^* a + v_i a^2 \quad &\text{if} \quad a \geqslant -y_i^*/(2v_i), \\ &= -y_i^{*2}/(4v_i) &\text{if} \quad a \leqslant -y_i^*/(2v_i). \end{aligned}$$

Notice that L_m is finite even if $y_i^* < 0$, in contrast to what we have seen when considering unmodified Lagrangians for problems involving inequality constraints. Since the constraint $y_i^* \geqslant 0$ is a nuisance when applying techniques involving gradients, consideration of (6.3.15) might be interesting from a computational point of view, even if the given problem is convex.

More abstractly, we could introduce modified bifunctions of the following type:

$$(6.3.16) \quad F_{\mathrm{m}}(x, y, z, v, w) = f(x) + \Sigma_i \tilde{p}_i(g_i(x), y_i, v_i) + \Sigma_j \tilde{q}_j(h_j(x), z_j, w_j)$$

at least if we restrict ourselves to Lagrangian duality. Here the functions $\tilde{p}_i$ and $\tilde{q}_j$ may take the value $+\infty$ so that terms containing indicator functions such as $\delta((x, y_i): G_i)$ with $G_i = \{x: g_i(x) \leqslant y_i\}$ could be taken up.

7
Some applications

7.1 Returning to the initial examples; a well-known inequality

Now that we are through the theory that we wanted to present, what can be said about the examples of chapter 1? Some of them have already received enough attention in the foregoing chapters: the norm problem of Example 1.2.1 is a simple case of Example 3.14.19, Examples 1.2.4, 1.2.5 and 1.2.9 are special cases of linear programming dealt with in 3.13 and Example 3.15.6, and results about the game of Example 1.2.7 can be found in chapter 6 for instance. Some other examples (1.2.2 about Chebyshev approximation, 1.2.6 about optimal control, and 1.2.8 about inventory replenishing) will be considered in subsequent sections. Hence the only one that remains is Example 1.2.3 involving a well-known inequality.

7.1.1 **Example 1.2.3.** The constraints of this example are linear and the objective function is concave. The latter follows from the fact that the Hessian H (the matrix of second-order partial derivatives) is negative semi-definite. We have that H is negative semi-definite if K defined by

$$K = \begin{bmatrix} a_1(a_1-1) & a_1 a_2 & \dots & a_1 a_n \\ a_2 a_1 & a_2(a_2-1) & \dots & a_2 a_n \\ \vdots & \vdots & & \vdots \\ a_n a_1 & a_n a_2 & \dots & a_n(a_n-1) \end{bmatrix}$$

is negative semi-definite. Adding to the first column all other columns of $K-\lambda I$ it follows from the fact that $a_1+\dots+a_n = 1$ that $\lambda = 0$ is a root of $\det(K-\lambda I) = 0$. Further, we have that $\det(K+a_i I)$ is equal to $\Pi_{j\neq i}(a_i-a_j)$ times a positive factor. Hence if

$$a_1 < a_2 < \dots < a_n \quad \text{then} \quad \operatorname{sgn}\det(K+a_i I) = (-1)^{n+i}.$$

It follows that $\det(K-\lambda I) = 0$ has n nonpositive solutions, and hence that K is negative semi-definite.

By the theory of Lagrangian duality we may now conclude that the required supremum is equal to

$$\sup\{\xi_1^{a_1}\ldots\xi_n^{a_n}+y_0^*(1-a_1\xi_1-\ldots-a_n\xi_n)+y_1^*\xi_1+\ldots+y_n^*\xi_n\}$$

for some $(y_0^*, y_1^*, \ldots, y_n^*) \in R_{n+1}$ with $(y_1^*, \ldots, y_n^*) \geqslant 0$ and that $y_i^*\xi_i = 0$ for $i = 1, \ldots, n$ for an optimal solution $x = (\xi_1, \ldots, \xi_n)$. But no ξ_i will be zero; hence $y_1^* = \ldots = y_n^* = 0$ and it easily follows by differentiation that all ξ_i must be equal and hence equal to 1.

7.2 Chebyshev approximation

In this section we consider the Chebyshev approximation problem (for the details of this, see Example 1.2.2). As shown in Chapter 1 this problem leads to an optimization problem without constraints. So how do we perturb the problem? All we can do is perturb the objective function F. There is no point in putting $F(x, y) = f(x)+g(y)$ where g is some function with $g(0) = 0$ (which is quite different from putting $F(x, y) = f(x)+g(x+y)$ as in Fenchel duality theory). But there is an alternative. For any function $h: [-1, +1] \to R$ let

$$|h| = \sup\{|h(t)|: -1 \leqslant t \leqslant +1\}.$$

Define F by

(7.2.1) $$F(x, y) = |q-y-\Sigma\xi_i(\cdot)^i| \quad \text{with} \quad y \in Y = C[-1, +1].$$

Here $C[-1, +1]$ is the space of all continuous functions on $[-1, +1]$ with $|y|$ as norm. By the Riesz representation theorem the dual of Y can be identified with the space of all regular bounded countably additive set functions μ defined on the Borel field of $[-1, +1]$, and for each $y^* \in Y^*$ there is a μ such that

(7.2.2) $$y^*y = \int_{-1}^{+1} y(t)\mu(\mathrm{d}t).$$

This implies that Y with its strong topology is compatible with Y^* with the σ-topology defined by the bilinear form y^*y given by (7.2.2).

It is easily shown that F is convex and closed as well as R-closed (for $F(x,y) \geqslant 0$) and that $\alpha = \inf_X F(x,0)$ is finite. Moreover, since we may use the strong topology of Y it is also easy to show that the perturbation function p is bounded above for y close to zero. Hence by Theorems 3.8.16, 3.8.11 and 3.8.1 the infimum of the primal problem is a minimum, the supremum of the dual problem is a maximum and the two are equal. Moreover by (4.4.18d) any primal solution x_o is an element of $\partial(-p^d)(0)$ and x_o can be found by differentiation from p^d if it is unique.

Let us now compute $\beta = \sup_{Y^*} \inf_{X,Y} (F(x,y) + y^* y)$. We put $r(t) = q(t) - \Sigma \xi_i t^i$ and replace y by $y + r$. Then we get that

$$\beta = \sup_{Y^*} \inf_X [\textstyle\int r(t)\, \mu(\mathrm{d}t) + \inf_Y \{|y| + \int y(t)\, \mu(\mathrm{d}t)\}],$$

where the integration is over $[-1, +1]$ and μ corresponds to y^* as in (7.2.2). When $|y^*|$ (which by definition is equal to

$$\sup \{|\textstyle\int y(t) \mu(\mathrm{d}t)| : |y| = 1\})$$

is larger than 1, then the expression between curly brackets can be made negative by a proper selection of y, and the infimum over Y is $-\infty$. On the other hand if $|y^*| \leqslant 1$ then this expression is nonnegative so that this infimum is zero. It follows that in order to compute the infimum over X we need only consider the term with r and take $|y^*| \leqslant 1$. Eliminating r by its defining equation we see that this infimum is only above $-\infty$ if $\int t^i \mu(\mathrm{d}t) = 0$ for all i, so that we obtain the following result.

$$(7.2.3) \qquad \beta = p^{\mathrm{d}}(0) = \sup \left\{ \int_{-1}^{+1} q(t) \mu(\mathrm{d}t) : |y^*| \leqslant 1, \right.$$
$$\left. \int_{-1}^{+1} t^i \mu(\mathrm{d}t) = 0,\ i = 0, \ldots, n \right\}.$$

If we want $p^{\mathrm{d}}(x^*)$ with $x^* = (\xi_0^*, \ldots, \xi_n^*) \in R_{n+1}$ for values of x^* other than zero, then we must replace $y^* y$ by $-x^* x + y^* y$ and the result is that we must replace the zero in the equality constraints of (7.2.3) by $-\xi_i^*$.

A simplification is obtained if we replace the interval $[-1, +1]$ by a finite subset $\{t_1, \ldots, t_m\}$. Putting $y_j^* = \mu(\{t_j\}), j = 1, \ldots, m$ we get from (7.2.3) that

(7.2.4) $\beta = \sup\{\Sigma_j q(t_j) y_j^* : \Sigma_j |y_j^*| \leqslant 1,$
$\Sigma_j t_j^i y_j^* = 0, i = 0, \ldots, n\},$

which is almost linear programming. If we knew the signs of the y_j^* it would be linear programming exactly. Since there are 2^m sign combinations, β is the maximum of 2^m numbers that could be obtained from the same number of linear programming problems. It can be shown that, since we have $n+2$ constraints (not counting the nonnegativity constraints $|y_j^*| \geqslant 0$) for each separate linear programming problem, there exists an optimal solution for y* having at most $n+2$ nonzero components. Hence the supremum of (7.2.4) is also achieved for a y^* with at most $n+2$ nonzero components. This result is well-known in the theory of Chebyshev approximation. It says that if the degree of the approximating polynomial is at most n then it is sufficient to select only $n+2$ out of the m elements of the set $\{t_1, \ldots, t_m\}$. The problem is which selection to make, and this is solved by special techniques.

We remark that Chebyshev approximation is not at all restricted to polynomial approximation and that $[-1, +1]$ may be replaced by any other finite closed interval.

7.3 A fixed time optimal control problem

As a simple example of a fixed time optimal control problem (see Example 1.2.6), let $X = C_n[0, 1]$ be the space of all continuous functions x: $[0, 1] \to R_n$ with norm

$$|x| = \sup\{|x(t)| : 0 \leqslant t \leqslant 1\}$$

and let $U = L_m^\infty[0, 1]$ be the space of all μ-essentially bounded μ-measurable functions u: $[0, 1] \to R_m$ with respect to the ordinary Lebesgue measure μ on $[0, 1]$ and with

$$|u| = \mu\text{-ess sup}\{|u(t)| : 0 \leqslant t \leqslant 1\}.$$

Further, let

$$\phi: R_n \times R_m \times [0, 1] \to R \quad \text{and} \quad \psi: R_n \times R_m \times [0, 1] \to R_n$$

be differentiable functions such that

(*a*) given $x(0)$ and the *control* u the *state* x can be solved from

$$x(t) = x(0) + \int_0^t \psi(x(\tau), u(\tau), \tau)\mathrm{d}\tau, \tag{7.3.1}$$

(*b*) q defined by

$$q(t) = k(t) - \int_0^t \{\psi_x(x(\tau), u(\tau), \tau)k(\tau) + \psi_u(x(\tau), u(\tau), \tau)v(\tau)\}\mathrm{d}\tau$$

belongs to X if $x, k \in X$ and $u, v \in U$,

(*c*) for all such q and all $\hat{u} \in U$ there exists an $\hat{x} \in X$ such that

$$\hat{x}(t) - \int_0^t \psi_x(x(\tau), u(\tau), \tau)\hat{x}(\tau)\mathrm{d}\tau = q(t).$$

The control is constrained by the condition that

$$u \in C = \{u: u \in U, -e \leqslant u(t) \leqslant +e \quad \text{if} \quad 0 \leqslant t \leqslant 1\}, \quad e = (1, 1, \ldots, 1), \tag{7.3.2}$$

that is, each component of $u(t)$ must lie between -1 and $+1$.

Instead of u the decision variable will be the pair (x, u) and (7.3.1) is considered as just another constraint. The problem is to find

$$\inf\left\{\int_0^1 \phi(x(\tau), u(\tau), \tau)\mathrm{d}\tau : u \in C, x \in X \text{ and } x \text{ satisfies (7.3.1)}\right\}. \tag{7.3.3}$$

We want to write this as $\inf\{f(x, u): h(x, u) = 0, u \in C\}$. The definition of f is simply

$$f(x, u) = \int_0^1 \phi(x(\tau), u(\tau), \tau)\mathrm{d}\tau \tag{7.3.4}$$

and that of $h(x, u): [0, 1] \to R_n$ is given by

$$h(x, u)(t) = x(t) - x(0) - \int_0^t \psi(x(\tau), u(\tau), \tau)\mathrm{d}\tau. \tag{7.3.5}$$

The computation of the Fréchet differentials of f and h is straightforward:

(7.3.6)

$$f'(x, u)(k, v) = \int_0^1 \{\phi_x(x(\tau), u(\tau), \tau)k(\tau) + \phi_u(x(\tau), u(\tau), \tau)v(\tau)\}\mathrm{d}\tau,$$

(7.3.7) $h'(x,u)(k,v)(t)$

$$= k(t) - \int_0^t \{\psi_x(x(\tau), u(\tau), \tau)k(\tau) + \psi_u(x(\tau), u(\tau), \tau)v(\tau)\}d\tau,$$

with $x, k \in X$ and $u, v \in U$.

We want to apply Theorem 5.4.3. By assumption (*b*) above, h' is continuous. In order to show that for some $\hat{x}$ and for some $\hat{u} \in \operatorname{int} C$ we have that $h'(x_o, u_o)((\hat{x}, \hat{u}) - (x_o, u_o)) = 0$ given an optimal solution (x_o, u_o), we take $\hat{u} = 0$; then $\hat{u} \in \operatorname{int} C$ because $U = L_m^\infty[0, 1]$. And we solve $\hat{x}$ from

$$\hat{x}(t) - \int_0^t \psi_x(x_o(\tau), u_o(\tau), \tau)\hat{x}(\tau)d\tau$$

$$= x_o(t) - \int_0^t \{\psi_x(x_o(\tau), u_o(\tau), \tau)x_o(\tau) + \psi_u(x_o(\tau), u_o(\tau), \tau)u_o(\tau)\}d\tau,$$

which is possible by assumption (*c*). This assumption is also sufficient for $h'(x_o, u_o)$ to be a mapping onto Y if we let $Y = X = C_n[0, 1]$.

Since Y is essentially the same space as in 7.2 its dual can again be identified with the space of all regular bounded countably additive set functions, but we prefer to represent Y^* now by the space $NBV[0, 1]$ of all normalized functions of bounded variation on the interval $[0, 1]$ which are continuous on the left and which vanish at $t = 1$. For each $y^* \in Y^*$ there exists a $\mu \in NBV[0, 1]$ such that y^*y can be represented by a Stieltjes integral

(7.3.8) $$y^*y = \int_0^1 y(t)d\mu(t),$$

where the integrand is the inner product in R_n of $y(t)$ and $d\mu(t)$.

According to (5.4.4) we may now conclude that if (x_o, u_o) is an optimal solution then for some $y_0^* \in Y^*$ we have that

(7.3.9)

$$(f'(x_o, u_o) + y_0^* h'(x_o, u_o))((x, u) - (x_o, u_o)) \geqslant 0 \quad \text{if} \quad u \in C.$$

Having first 'packed' the original formulation in terms of a Lagrangian duality problem, let us now 'unpack' (7.3.9) and see what comes out. First we take $u = u_o$ and $x = k + x_o$ with $k \in X$. Since (7.3.9) does not put any restriction on x, we can take any $k \in X$. We take a sequence of functions k which converge to the function that is equal to 1 for all τ, $0 \leqslant \tau \leqslant t$ and that is equal to 0 if $t < \tau \leqslant 1$, for some t. After some calculations which include an integration by parts we get

$$\int_0^t \phi_x(x_o(\tau), u_o(\tau), \tau)d\tau + \int_0^t d\mu_o(\tau) + \int_0^t \mu_o(\tau)\psi_x(x_o(\tau), u_o(\tau), \tau)d\tau = 0,$$

from which it follows that μ_o is differentiable, and that if $0 \leqslant t \leqslant 1$,

$$\phi_x(x_o(t), u_o(t), t) + d\mu_o(t)/dt + \mu_o(t)\psi_x(x_o(t), u_o(t), t) = 0. \tag{7.3.10}$$

We now take $x = x_o$ and obtain from (7.3.9) that if $v = u - u_o \in -u_o + C$

$$\int_0^1 \phi_u(x_o(t), u_o(t), t)v(t)dt + \int_0^1 \mu_o(t)\psi_u(x_o(t), u_o(t), t)v(t)dt \geqslant 0. \tag{7.3.11}$$

These results prompt us to define the function

$$H: R_n \times R_m \times [0, 1] \times R_n \to R,$$

called the *Hamiltonian*,

$$H(\tilde{x}, \tilde{u}, t, \tilde{\mu}) = \phi(\tilde{x}, \tilde{u}, t) + \tilde{\mu}\psi(\tilde{x}, \tilde{u}, t) \tag{7.3.12}$$

with $\tilde{x}$, $\tilde{\mu} \in R_n$, $\tilde{u} \in R_m$ and $t \in [0, 1]$. Then (7.3.11) becomes

$$\int_0^1 H_u(x_o(t), u_o(t), t, \mu_o(t))v(t)dt \geqslant 0 \quad \text{if} \quad v + u_o \in C. \tag{7.3.13}$$

We claim that from this it follows that

$$H(x_o(t), u_o(t), t, \mu_o(t)) \leqslant H(x_o(t), u(t), t, \mu_o(t)) \quad \text{if} \quad u \in C \quad \text{and} \quad t \in [0, 1]. \tag{7.3.14}$$

This is the so-called *minimum principle* of our control problem. Notice that the values of x, t and μ are the same on both sides of (7.3.14); only u has different values.

To see that (7.3.14) is correct first notice that

$$L_m^\infty[0,1] \subset L_m^1[0,1] \quad \text{where} \quad L_m^1[0,1]$$

is the space of all absolutely integrable functions u on $[0, 1]$ with norm $|u|_1 = \int_0^1 |u(t)|\mathrm{d}t$. For clarity we put $|u|_\infty = |u|$.

Now suppose (7.3.14) were not true. Then

$$H(x_o(t), u_o(t), t, \mu_o(t)) \geqslant \delta + H(x_o(t), u'(t), t, \mu_o(t))$$

for some $\delta > 0$, some $u' \in C$ and all t in a neighbourhood W of some t'. For $\epsilon > 0$ let $W(\epsilon)$ be the ϵ-neighbourhood of t' and define u_ϵ by

$$u_\epsilon(t) = u_o(\mathrm{t}) \quad \text{if} \quad t \notin W(\epsilon),$$
$$u_\epsilon(t) = u'(t) \quad \text{if} \quad t \in W(\epsilon).$$

Then $u_\epsilon \in C$ and $|u_\epsilon - u_o|_1 \leqslant |u_\epsilon - u_o|_\infty \epsilon \leqslant 2\epsilon$, so that

$$-\delta\epsilon \geqslant \int_0^1 \{H(x_o(t), u_\epsilon(t), t, \mu_o(t)) - H(x_o(t), u_o(t), t, \mu_o(t))\}\mathrm{d}t$$
$$= \int_0^1 H_u(x_o(t), u_o(t), t, \mu_o(t))(u_\epsilon(t) - u_o(t))\mathrm{d}t + o(\epsilon).$$

In view of (7.3.13) this is at least $o(\epsilon)$, so that $-\delta\epsilon \geqslant o(\epsilon)$, which cannot be true.

Summarizing we find that for some function $\mu_o\colon [0, 1] \to R$

(7.3.15) $$\mathrm{d}\mu_o(t)/\mathrm{d}t = -H_x(x_o(t), u_o(t), t, \mu_o(t)), \quad 0 \leqslant t \leqslant 1; \quad \mu_o(1) = 0$$

and

(7.3.16) $$H(x_o(t), u_o(t), t, \mu_o(t)) \leqslant H(x_o(t), u(t), t, \mu_o(t)), \quad 0 \leqslant t \leqslant 1, u \in C.$$

Together with (7.3.1) these equations and inequalities are in general sufficient to compute (x_o, u_o) as well as μ_o. Notice that (7.3.15) is a differential equation for μ_o with the value of $\mu_o(t)$ fixed for the *final* t, namely $t = 1$, whereas in (7.3.1) the *initial* value $x(0)$ is given. This situation is typical in optimal control and may result in certain numeric problems when solving the entire system of equations.

Equation (7.3.15) is called the *adjoint equation* and the variable μ_o the *adjoint variable.*

There exists a close relationship between the Hamiltonian and the Lagrangian which reads (with μ taking the place of y^*)

$$(7.3.17)\qquad \begin{aligned} &L(x, u, \mu)\\ &= \int_0^1 \phi(x(t), u(t), t)\mathrm{d}t + \int_0^1 (x(t)-x(0))\mathrm{d}\mu(t)\\ &\quad - \int_0^1 \int_0^t \psi(x(\tau), u(\tau), \tau)\mathrm{d}\tau\mathrm{d}\mu(t), \end{aligned}$$

so that

$$(7.3.18)\qquad L(x, u, \mu) = \int_0^1 (x(t)-x(0))\mathrm{d}\mu(t) + \int_0^1 H(x(t), u(t), t, \mu(t))\mathrm{d}t,$$

and ignoring the first term on the right we see from this that roughly speaking 'the Hamiltonian is the integrand of the Lagrangian'.

While we are discussing the Lagrangian, notice that only the constraint (7.3.1) is incorporated in it and that the constraint $u \in C$ is left out, so that the latter constraint appears again in (7.3.9). In the present case, however, we could have incorporated all constraints in L. Then we should have added the terms

$$\int_0^1 (-e-u(t))\nu^{(1)}(t)\mathrm{d}t \quad \text{and} \quad \int_0^1 (-e+u(t))\nu^{(2)}(t)\mathrm{d}t \qquad \text{with} \quad \nu^{(1)}, \nu^{(2)} \geqslant 0.$$

Instead of the minimum principle (7.3.16) we would obtain

$$(7.3.19)\qquad \nu_o^{(1)}(t) - \nu_o^{(2)}(t) = H_u(x_o(t), u_o(t), t, \mu_o(t))$$

together with conditions of the type $y_o^* g(x_o) = 0$ namely

$$(7.3.20)\quad (e+u_o(t))\nu_o^{(1)}(t) = (-e+u_o(t))\nu_o^{(2)}(t) = 0, \quad 0 \leqslant t \leqslant 1.$$

Equations (7.3.15) and (7.3.19) together are called the *Hamiltonian equations*, in particular if the constraint $u \in C$ is absent so that (7.3.19) reduces to $H_u(\ldots) = 0$, which can also be obtained from the minimum principle (7.3.16).

It is also possible to specify the second-order conditions for the present problem, but this we leave to the reader.

Finally we would like to remark that the reasoning of this section is not at all restricted to the simple problem we have taken. Problems such as those where restrictions are put on the final state $x(1)$ (in which case the conditions of the type $y_0^* g(x_o) = 0$ lead to so-called *transversality conditions*), or those involving partial differential state equations, can in principle be treated within the framework of the theory of optimization as developed in this book; that is, from the standpoint of mathematical programming. This means that mathematical programming leads to an important alternative to the classical *calculus of variations*.

One might object that we have confined ourselves to *open loop control*, because the state x is determined by the control u (via equation (7.3.1)) whereas conversely u is not influenced by x, or in other words x is not *fed back* to u. In *closed loop control*, however, $u(t)$ is made dependent on $x(\tau)$ for all τ, $0 \leqslant \tau \leqslant t$; that is, the control u is made dependent on the past and present values of the state x. Defining for each t the function $\bar{x}_t$: $[0, t] \to R_n$ by $\bar{x}_t(\tau) = x(\tau)$ for $0 \leqslant \tau \leqslant t$, we can express the *feedback* by putting $u(t) = r(\bar{x}_t, t)$. This leads to the problem of finding

$$\inf\left\{\int_0^1 \phi(x(t), r(\bar{x}_t, t), t)\mathrm{d}t: \right.$$

$$x(t) = x(0) + \int_0^1 \psi(x(t), r(\bar{x}_t, t), t)\mathrm{d}t,$$

$$\left. \bar{x}_t(\tau) = x(\tau),\ 0 \leqslant \tau \leqslant t,\ u(t) = r(\bar{x}_t, t),\ u \in C \right\}$$

by the proper selection of (x, r) rather than (x, u). Although more complicated than the problem of finding the infimum of (7.3.3) (notice that while $x(t)$, $r(\bar{x}_t, t)$ and t are finite-dimensional, $\bar{x}_t$ is an element of $C_n[0, t]$), this problem can in principle be tackled by mathematical programming methods. In particular this is true if we need to know only $\bar{x}_t(t) = x(t)$ rather than $\bar{x}_t(\tau)$ for all $\tau \leqslant t$; that is, when the system '*has no memory*', which is often the case.

7.4 A two-period stochastic inventory problem

The details of the problem we want to solve in this section are given in Example 1.2.8. The restrictions to only two periods, to linear cost functions, to identical data for each period, and to only one commodity are not essential, so that the model can be extended considerably. As should be clear from the description given in Example 1.2.8 we assume that '*backlogging*' occurs, which means that unsatisfied demand will be satisfied later on, with the exception, however, of the last period (although models exist where the shortage accumulated in the last period is satisfied at the end of that period). Strictly speaking it is possible that no demand at all is satisfied, so that all demand accumulates.

For simplicity we assume further that the probability distribution for the demand is given by a density $\phi(s)$ (so that $\mu(\mathrm{d}s) = \phi(s)\mathrm{d}s$), and that $S = [0, 1]$.

As with the Chebyshev approximation problem (see 7.2) we have to perturb the objective function. This time we perturb the demand. Let $y_1(s)$ be the decrease of the demand in the first period and let $y_2(s, s')$ be the decrease of the demand s' of the second period provided the demand in the first period is s. The reason why we make y_2 depend on s will become clear soon.

The bilinear forms x^*x and y^*y will be as follows:

(7.4.1)

$$x^*x = x_1^*x_1 + x_2^*x_2 \quad \text{with} \quad x_2^*x_2 = \int_0^1 x_2^*(s)x_2(s)\phi(s)\mathrm{d}s$$

and

(7.4.2)
$$y^*y = \int_0^1 y_1^*(s)y_1(s)\phi(s)\mathrm{d}s + \int_0^1\int_0^1 y_2^*(s,s')y_2(s,s')\phi(s')\,\phi(s)\,\mathrm{d}s'\,\mathrm{d}s.$$

As $F(x, 0)$ is given at the end of Example 1.2.8 it is now easy to calculate $F(x, y) - x^*x + y^*y$. But before writing down the result we make an important substitution:

(7.4.3)
$$v_1(s) = x_1 - s + y_1(s),$$
$$v_2(s, s') = v_1(s) + x_2(s) - s' + y_2(s, s').$$

Obviously, we can compute y_1 and y_2 from (7.4.3) for arbitrary functions v_1 and v_2, taking the remaining variables fixed. But this is only possible if y_2 depends on both s and s', as we assumed. The point is that in the subsequent analysis we can let v_1 and v_2 vary independently of each other. We have

(7.4.4)

$$\begin{aligned} F(x,y)-x^*x+y^*y \\ &= cx_1+\int_0^1 \{cx_2(s)+h(v_1(s))^+ + u(-v_1(s))^+\}\phi(s)\mathrm{d}s \\ &\quad+\int_0^1\int_0^1 \{h(v_2(s,s'))^+ + u(-v_2(s,s'))^+\}\phi(s')\phi(s)\mathrm{d}s'\,\mathrm{d}s \\ &\quad-x_1^*x_1-\int_0^1 x_2^*(s)x_2(s)\phi(s)\mathrm{d}s \\ &\quad+\int_1^0 y_1^*(s)\{v_1(s)-x_1+s\}\phi(s)\mathrm{d}s \\ &\quad+\int_0^1\int_0^1 y_2^*(s,s')\{-v_1(s)+v_2(s,s')-x_2(s)+s'\} \\ &\qquad\times\phi(s')\phi(s)\mathrm{d}s'\,\mathrm{d}s, \end{aligned}$$

at least if $x \geqslant 0$, because $F(x,y)=+\infty$ if $x \ngeq 0$. Rather than computing $\sup_{y^*}\inf_{X,Y}$ of $F(x,y)-x^*x+y^*y$, we compute $\sup_{Y^*}\inf_{X,V}$ where V is the space of vs and $v=(v_1,v_2)$. In view of the fact that we must compute a supremum over Y^* we can restrict y^* by the following constraints,

(7.4.5a) $$c-x_1^*-\int_0^1 y_1^*(x)\phi(s)\mathrm{d}s \geqslant 0,$$

(7.4.5b) $$c-x_2^*(s)-\int_0^1 y_2^*(s,s')\phi(s')\mathrm{d}s' \geqslant 0,$$

(7.4.6a) $$-h \leqslant y_1^*(s)-\int_0^1 y_2^*(s,s')\phi(s')\mathrm{d}s' \leqslant u,$$

(7.4.6b) $$-h \leqslant y_2^*(s,s') \leqslant u,$$

because otherwise the infimum over X and V will be $-\infty$. Conditions (7.4.5) are fairly obvious, and to see that we may require conditions (7.4.6) to be true observe that e.g. $h(v_1(s))^+ + u(-v_1(s))^+$ is the maximum of two linear functions (linear in $v_1(s)$, that is) so that if (7.4.6a) is violated the infimum over v_1 is $-\infty$.

Given that (7.4.5) and (7.4.6) hold it is easy to verify that the infimum over X and V is achieved at $x = 0$ and $v = 0$, and all that is left on the right-hand side of (7.4.4) is

(7.4.7) $$\int_0^1 y_1^*(s)s\phi(s)\mathrm{d}s + \int_0^1\int_0^1 y_2^*(s,s')s'\phi(s')\phi(s)\mathrm{d}s'\,\mathrm{d}s.$$

Hence we must calculate the supremum of (7.4.7) subject to (7.4.5) and (7.4.6). This is perhaps best done by substituting new variables for the integrals in (7.4.5) and (7.4.6), because if we fix these integrals then in (7.4.6*a*) we 'uncouple' y_1 and y_2 so that we are able to compute the (so restricted) supremum of (7.4.7) as the sum of two suprema. The most convenient choice of the new variables is as follows.

(7.4.8*a*)
$$(h+u)k_1 = u - \int_0^1 y_1^*(s)\phi(s)\mathrm{d}s + \int_0^1\int_0^1 y_2^*(s,s')\phi(s')\phi(s)\mathrm{d}s'\,\mathrm{d}s,$$

(7.4.8*b*) $$(h+u)k_2(s) = u - \int_0^1 y_2^*(s,s')\phi(s')\mathrm{d}s'.$$

Assume that $h+u > 0$, which is quite a reasonable assumption; then (7.4.5) leads to

(7.4.9)
$$k_1 + \int_0^1 k_2(s)\phi(s)\mathrm{d}s \geqslant \frac{2u-c+x_1^*}{h+u}; \quad k_2(s) \geqslant \frac{u-c+x_2^*(s)}{h+u}.$$

And integration of (7.4.6) leads to

(7.4.10) $$0 \leqslant k_1 \leqslant 1; \quad 0 \leqslant k_2(s) \leqslant 1,$$

which shows why we made this particular choice of k_1 and k_2.

Suppose we fix k_1 and k_2 so as to satisfy (7.4.5) and (7.4.10); then the supremum over y_1^* and y_2^* is achieved at y_1^* and y_2^* satisfying

(7.4.11) $y_1^*(s)$ is equal to its lower bound,
i.e. $-h+u-(h+u)k_2(s)$, if $0 \leqslant s < s_0$,

$y_1^*(s)$ is equal to its upper bound,
i.e. $2u-(h+u)k_2(s)$, if $s_0 < s \leqslant 1$,

$y_2^*(s,s')$ is equal to its lower bound,
i.e. $-h$, if $0 \leqslant s' < s_0'(s)$,

$y_2^*(s,s')$ is equal to its upper bound,
i.e. u, if $s_0'(s) < s' \leqslant 1$,

where s_0 and s_0' are defined by

$$(7.4.12) \quad \int_0^{s_0} \phi(s)\mathrm{d}s = k_1, \quad \text{and} \quad \int_0^{s'_0(s)} \phi(s')\mathrm{d}s' = k_2(s).$$

The existence of s_0 and $s_0'(s)$ is ensured by condition (7.4.10). The bounds for y_2^* are obtained from (7.4.6*b*) and those for y_1^* from (7.4.6*a*) and (7.4.8*b*).

We can now completely eliminate y^*, and it is convenient to eliminate k_1 and $k_2(s)$ as well. Then a straightforward calculation, where integrals over $(r, 1)$ are written as the difference of integrals over $(0, 1)$ and $(0, r)$, leads to the following optimization problem, where $\bar{s}$ is the expectation of s: find

(7.4.13)

$$3u\bar{s} - (h+u)\inf\left\{\int_0^{s_0} s\phi(s)\mathrm{d}s + \int_0^1\int_0^{s'_0(s)} (s+s')\phi(s')\phi(s)\mathrm{d}s'\mathrm{d}s,\right.$$

$$\text{subject to} \qquad 0 \leqslant s_0, \quad s_0'(s) \leqslant 1,$$

$$\int_0^{s_0} \phi(s)\mathrm{d}s + \int_0^1\int_0^{s'_0(s)} \phi(s')\phi(s)\mathrm{d}s'\mathrm{d}s \geqslant \frac{2u-c+x_1^*}{h+u},$$

$$\left.\int_0^{s'_0(s)} \phi(s')\mathrm{d}s \geqslant \frac{u-c+x_2^*(s)}{h+u}\right\}.$$

Assume that $u > c > 0$ and $h > 0$, which implies the earlier assumption that $h+u > 0$, and that x^* is sufficiently small. Assume further that $\phi(s)$ is zero only at isolated points. Then there exist unique $t'(s)$ and t satisfying

$$(7.4.14) \quad \int_0^{t'(s)} \phi(s')\mathrm{d}s' = \frac{u-c+x_2^*(s)}{h+u}, \quad 0 < t'(s) < 1,$$

and

(7.4.15)
$$\int_0^t \phi(s)\mathrm{d}s + \int_0^1 \int_0^{m(s)} \phi(s')\phi(s)\mathrm{d}s'\mathrm{d}s = \frac{2u-c+x_1^*}{h+u}, \quad 0 < t < 1,$$

where

(7.4.16)
$$m(s) = \max(t-s, t'(s)).$$

The existence of $t'(s)$ is easy to show, and to see that t exists observe that the left-hand side of (7.4.15) is smaller or larger than the right-hand side if $t = 0$ or $t = 1$, respectively; for if $t = 0$ then $m(s) = t'(s)$, and if $t = 1$ then $m(s) \geqslant t'(s)$. The uniqueness of $t'(s)$ and t follows from the assumption that $\phi(s)$ is zero only at isolated points. Since the left-hand side of (7.4.15) is strictly increasing with increasing t it follows from $c > 0$ that $t'(s) < t$, so that

(7.4.17)
$$0 < t'(s) < t < 1.$$

Trivially $s_0'(s) \geqslant t'(s)$ (because of the last constraint of (7.4.13)) and if $x_2^* = 0$ then $t'(s)$ is constant, in which case we put $t' = t'(s)$.

We shall now solve (7.4.13). Let $\mathrm{d}s_0$ and $\mathrm{d}s_0'$ be small changes of s_0 and s_0'; then the left-hand side of the second last constraint changes by

(7.4.18)
$$\mathrm{d}g = \phi(s_0)\mathrm{d}s_0 + \int_0^1 \phi(s_0'(s))\phi(s)\mathrm{d}s_0'(s)\mathrm{d}s$$

and the objective function by

(7.4.19)
$$\mathrm{d}f = s_0\phi(s_0)\mathrm{d}s_0 + \int_0^1 (s+s_0'(s))\phi(s_0'(s))\phi(s)\mathrm{d}s_0'(s)\mathrm{d}s.$$

We claim that at optimum $s_0 = t$. For when $s_0 < t$ it is not possible that $s_0'(s) \leqslant m(s)$ everywhere, because then the left-hand side of the second last constraint would be smaller than the left-hand side of (7.4.15), so that this constraint would be violated. So $s_0'(s) > m(s)$ somewhere, and $s_0'(s) > t'(s)$ there. This means that we can take $\mathrm{d}s_0'(s) < 0$ there and $\mathrm{d}s_0'(s) = 0$ elsewhere, and $\mathrm{d}s_0 > 0$ such that $\mathrm{d}g = 0$ and such that no constraint is violated, for $s_0 < 1$ and $s_0'(s) > t'(s) > 0$ there. Consequently, as we started from an optimal solution, $\mathrm{d}f \geqslant 0$, hence $\mathrm{d}f - s_0\mathrm{d}g \geqslant 0$, from which it is seen that

$$-s_0 + s_0'(s) + s \leqslant 0, \quad \text{so that} \quad t'(s) < s_0'(s) \leqslant s_0 - s < t - s.$$

Hence $m(s) = t-s > s_0'(s)$, contradicting $s_0'(s) > m(s)$. And when $s_0 > t$ it is not possible that $s_0'(s) = 1$ everywhere, because then the last two constraints would be satisfied with strict inequality, which is certainly not optimal. So $s_0'(s) < 1$ somewhere. Now we can take $ds_o'(s) > 0$ there and $ds_o'(s) = 0$ elsewhere, and $ds_o < 0$ such that $dg = 0$ and we would have $-s_0 + s + s_0'(s) \geqslant 0$; hence $1 > s_0'(s) \geqslant s_0 - s > t-s$ or $t > 1$, which is impossible. We have found, therefore, that at optimum $s_0 = t$.

Knowing this it is now easy to show that at optimum $s_0'(s) = m(s)$. For take $dg = 0$ and $ds_o'(s) > 0$ everywhere, then it follows as before that $s_0'(s) \geqslant t-s$. And since $s_0'(s) \geqslant t(s)$ always, we have $s_0'(s) \geqslant m(s)$ everywhere, and $s_0'(s) = m(s)$, because $s_0'(s) > m(s)$ would not be optimal.

It would seem then that we had already found the optimal solution, when we defined $t'(s)$, t and $m(s)$.

Having solved the perturbed dual problem, that is having found $p^d(x^*)$, at least in principle, we now want to jump back to the primal problem and compute x_o from $x_o \in \partial(-p^d)(0)$. It so happens that in the present case we can obtain x_o by Fréchet differentiation. After a not too tedious calculation we find that the optimal solution for x_1 is t, and after some more calculations that the optimal solution for x_2 is given by $x_2(s) = \max(0, s-(t-t'))$. This is the well-known solution usually obtained by means of *dynamic programming*. It says that the production in the first period should be t, and that if at the beginning of the second period the inventory, i.e. $t-s$, is smaller than t' then the production should be $s-t+t'$ so that the inventory is increased to t', whereas if $t-s > t'$ no production should take place in the second period.

Summarizing, we see that the solution of the dual problem when reduced to that of finding (7.4.13) is

(7.4.20) $$s_0 = t \quad \text{and} \quad s_0'(s) = m(s)$$

and that the primal solution is

(7.4.21) $$x_1 = t \quad \text{and} \quad x_2(s) = \max(0, s-(t-t')),$$

at least if $u > c > 0$, $h > 0$ and $\phi(s) = 0$ only at isolated points of $[0, 1]$. Notice that $t' - m(s) = \min(0, s-(t-t'))$.

The reader will have seen that we have not bothered about the precise choice of the relevant spaces and the truth of duality results in this example. This time it is left to him.

7.5 Theorems of the alternative; multi-objective optimization

In a *theorem of the alternative* (sometimes called *transposition theorem*) two sets, say Q and R, play a part, and the theorem itself takes the following form:

(7.5.1)
$$\text{if } Q = \varnothing \text{ then } R \neq \varnothing,$$
$$\text{if } Q \neq \varnothing \text{ then } R = \varnothing;$$

hence exactly one of the two sets must be empty.

We have already come across such a theorem, i.e. Farkas' lemma (Theorem 5.2.3). For let $c, x \in R_n$, $y^* \in R_m$, let A be an $m \times n$ matrix and let

(7.5.2)
$$Q = \{x: cx < 0, Ax \leqslant 0\},$$
$$R = \{y^*: y^* \geqslant 0, c + y^*A = 0\}.$$

Then if $Ax \leqslant 0$ implies that $cx \geqslant 0$ we obviously have that $Q = \varnothing$, so that if it followed from this that $R \neq \varnothing$ we would indeed have that $c + y^*A = 0$ for some $y^* \geqslant 0$. One might say that Theorem 5.2.3 does not claim anything when $Q \neq \varnothing$, but to conclude from this that $R = \varnothing$ is trivial, since $Q \neq \varnothing$ and $R \neq \varnothing$ immediately leads to a contradiction.

In what follows the constraint $Ax \leqslant 0$ is augmented with $Bx = 0$, and the inner product cx is replaced by the matrix product Cx, although x is still in R_n. The latter means that the objective function cx is replaced by a number of objective functions $c_j x$, where c_j is the jth row of C. We come back to this at the end of this section. Now the simple question arises of how to replace the inequality $cx < 0$ occurring in the definition of Q in (7.5.2). We shall consider three possibilities: (*a*) $Cx < 0$, (*b*) $Cx \leqslant 0$, $Cx \neq 0$, and (*c*) a combination of (*a*) and (*b*).

7.5.3 **Theorem** (Motzkin). Let $x \in R_n$ and $y^* \in R_m$, and let C, B and A be matrices of appropriate sizes, such that C has at least one row and at least one column. Then either

(7.5.4)
$$Q = \{x: Cx < 0, Ax \leqslant 0, Bx = 0\}$$

is nonempty, or

$$(7.5.5)\quad R = \{y^*: y^* = (y_1^*, y_2^*, y_3^*), y_1^* C + y_2^* A + y_3^* B = 0,\\ y_1^* \geqslant 0, y_1^* \neq 0, y_2^* \geqslant 0\}$$

is nonempty, but not both.

Proofs. In all three proofs that follow we shall only show that $Q = \emptyset$ implies that $R \neq \emptyset$, since it is trivial to verify that $Q \neq \emptyset$ and $R \neq \emptyset$ cannot go together.

Proof 1. Let

$$V = \{y_1: y_1 > Cx \text{ for some } x \text{ such that } Ax \leqslant 0,\ Bx = 0\}.$$

Then $Q = \emptyset$ means that $0 \notin V$, so that we can separate $\{0\}$ from the convex set V. Hence for some $y_1^* \neq 0$ we have $y_1^* y_1 \geqslant 0$ if $y_1 \in V$. Put $y_1 = Cx + z$ with x such that $Ax \leqslant 0$, $Bx = 0$ and $z > 0$. Then $y_1^* Cx + y_1^* z \geqslant 0$, or $y_1^* Cx \geqslant -y_1^* z$, so that $y_1^* Cx \geqslant 0$ for all x such that $Ax \leqslant 0$ and $Bx = 0$. Taking $x = 0$ we find that $y_1^* z \geqslant 0$ for all $z > 0$; hence $y_1^* \geqslant 0$. It also follows that $0 = \inf\{y_1^* Cx: Ax \leqslant 0, Bx = 0\}$, which is linear programming, and from Theorem 3.13.8 that $y_1^* C + y_2^* A + y_3^* B = 0$ for some $y_2^* \geqslant 0$ and some y_3^*, which shows that R is not empty.

Proof 2. Instead of invoking duality theory of linear programming in the last part of Proof 1 we can rely on Farkas' lemma by putting

$$Q' = \{x: y_1^* Cx < 0, Ax \leqslant 0, Bx = 0\}$$

and $$R' = \{(y_2^*, y_3^*): y_1^* C + y_2^* A + y_3^* B = 0, y_2^* \geqslant 0\}.$$

Proof 3. A third proof is obtained by using the ideas of the proof of Theorem 3.13.8 (duality of linear programming), and applying the result of Appendix D. Observe that because the equality in the definition of R is homogeneous in y^*, we can equally well require that $y_1^* \geqslant 0$ and $y_{1i}^* = 1$ for some i, where y_{1i}^* is the ith component of y_1^*. Now define G and H by

(7.5.6)

$$G = \{x^*: x^* = y_2^* A + y_3^* B \text{ for some } y_2^* \geqslant 0 \text{ and some } y_3^*\},$$
$$H = \{x^*: x^* = -y_1^* C \text{ for some } y_1^* \geqslant 0 \text{ with } y_{1i}^* = 1 \text{ for some } i\}.$$

Then we have to show that $Q = \emptyset$ implies that $G \cap H \neq \emptyset$. Suppose to the contrary that $G \cap H = \emptyset$. Since both G and H are convex and closed (the latter follows from Appendix D) they can be separated strongly. Hence for some x_o and some $\delta > 0$,

$$(y_1^* C + y_2^* A + y_3^* B) x_o \leqslant -\delta$$

if (y_1^*, y_2^*, y_3^*) is restricted as indicated. Then necessarily $Ax_o \leqslant 0$, $Bx_o = 0$ and $Cx_o \leqslant 0$, but $Ax_o \leqslant 0$ and $Bx_o = 0$ means that $Cx_o \nless 0$, because Q is empty; hence $y_1^* Cx_o = 0$ for a suitable y_1^*. Taking $y_2^* = 0$ and $y_3^* = 0$ we arrive at $0 \leqslant -\delta$ which is false.

Notice that Farkas' lemma is a special case of Motzkin's theorem.

Let us now replace the condition $Cx < 0$ by $Cx \leqslant 0$, $Cx \neq 0$.

7.5.7 **Theorem** (Tucker). Let $x \in R_n$ and $y^* \in R_m$, and let C, B and A be matrices of appropriate sizes, such that C has at least one row and at least one column. Then either

(7.5.8) $$Q = \{x \colon Cx \leqslant 0, Cx \neq 0, Ax \leqslant 0, Bx = 0\}$$

is nonempty, or

(7.5.9) $$R = \{y^* \colon y^* = (y_1^*, y_2^*, y_3^*), y_1^* C + y_2^* A + y_3^* B = 0,\quad y_1^* > 0, y_2^* \geqslant 0\}$$

is nonempty, but not both.

Proof. Reducing the theorem to either linear programming or to Farkas' lemma does not seem to be obvious, so let us proceed as in the third proof above. Now we may require that $y_1^* \geqslant e$ with $e = (1, \ldots, 1)$ instead of $y_1^* > 0$. So let

(7.5.10)

$$G = \{x^* \colon x^* = y_2^* A + y_3^* B \text{ for some } y_2^* \geqslant 0 \text{ and some } y_3^*\}$$
$$H = \{x^* \colon x^* = -y_1^* C \text{ for some } y_1^* \geqslant e\}.$$

Again we must show that $G \cap H \neq \varnothing$ if $Q = \varnothing$. If this were not true then $(y_1^* C + y_2^* A + y_3^* B) x_o \leqslant -\delta$ for some x_o and some $\delta > 0$. Again $Ax_o \leqslant 0$ and $Bx_o = 0$, so that either $Cx_o = 0$ or $(Cx_o)_i > 0$ for some i, because Q is empty. But also $Cx_o \leqslant 0$, so that $Cx_o = 0$, which as before leads to a contradiction.

Special cases of Theorems 7.5.3 and 7.5.7 are obtained by deleting the constraints $Ax \leqslant 0$ and $Bx = 0$, as well as y_2^* and y_3^*. Then

Motzkin's theorem reduces to a theorem due to Gordan (1873), and Tucker's to one due to Stiemke (1915). (The date for Farkas' lemma is 1902.)

The next theorem is obtained by combining the preceding two.

7.5.11 **Theorem** (Slater). Let $x \in R_n$ and $y^* \in R_m$, and let C, C', B and A be matrices of appropriate sizes, such that at least one of C and C' has at least one row and at least one column. Then either

(7.5.12) $Q = \{x\colon Cx < 0, C'x \leqslant 0, C'x \neq 0, Ax \leqslant 0, Bx = 0\}$
is nonempty, or

(7.5.13)
$$R = \{y^*\colon y^* = (y_1^*, y_1'^*, y_2^*, y_3^*), y_1^* C + y_1'^* C' + y_2^* A + y_3^* B = 0,$$
$$y_1^* \geqslant 0, y_1'^* \geqslant 0, y_2^* \geqslant 0, \text{ and } y_1^* \neq 0 \text{ or } y_1'^* > 0\}$$
is nonempty, but not both.

Proof. The proof is left to the reader.

Apart from the proofs given, the theorems of the alternative considered here can also be shown by an argument that is inductive (with respect to the size of the matrices involved) and that does not rely on separation. This approach is perhaps more 'elementary' but not very easy. In this connection it is worth mentioning that the duality results of linear programming can also be obtained by an inductive proof. In fact this proof can be formulated constructively in such a way that the primal as well as the dual solutions can be effectively obtained from it, resulting in the well-known *simplex method* and its variants.

The theorems of this section invite us to generalize the idea of the optimization problem. Instead of dealing with a single objective cx, we could be dealing with a number of them, say $c_1 x, c_2 x, \ldots, c_g x$. Let us define C as the matrix whose rows are $c_1, c_2, \ldots, c_g$. Then the generalized problem is that of finding

(7.5.14) $$\text{'inf'}\{Cx\colon Ax \leqslant 0, Bx = 0\},$$

which must be understood as follows. Find an x_0 such that x_0 is feasible, i.e. such that $Ax_0 \leqslant 0$, $Bx_0 = 0$, and such that if x is any feasible solution $C(x-x_o) < 0$ does *not* hold. Or, instead, find a feasible x_o such that if x is any feasible solution, $C(x-x_o) \leqslant 0$, $C(x-x_o) \neq 0$ does *not* hold. The former interpretation is derived from Motzkin's theorem, the latter from Tucker's. Of course we can also consider a mixture of the two just as in Slater's theorem. In the case where $C(x-x_o) \leqslant 0$, $C(x-x_o) \neq 0$ one speaks of *Pareto optimality*. Then x_o is optimal if there is *no* feasible x which is at least as good as far as each separate objective is concerned and moreover is better for some of the objectives.

Theorems of the alternative and multi-objective optimization are not limited to linear problems. For let $F: X \to R_g$, $g: X \to Y$ be convex functions, let $h: X \to Z$ be an affine function, and let C be a convex set. Then a typical alternative is obtained by putting

$$(7.5.15)\quad \begin{aligned} Q &= \{x: F(x) < 0, g(x) \leqslant 0, h(x) = 0, x \in C\} \\ R &= \{(r^*, y^*, z^*): r^*F(x) + y^*g(x) + z^*h(x) \geqslant 0 \\ &\qquad \text{for all } x \in C, \text{ and } r^* \geqslant 0, r^* \neq 0, y^* \geqslant 0\}. \end{aligned}$$

Again Q and R cannot both be nonempty. And if Q is empty we can follow the first proof of Theorem 7.5.3 to arrive at the ordinary optimization problem of finding

$$\inf\{r^*F(x): g(x) \leqslant 0, h(x) = 0, x \in C\}.$$

If we assume that certain regularity conditions hold it can be shown from this that R is nonempty. We leave it to the reader to select the appropriate conditions as well as the appropriate duality theorems developed earlier in this book.

7.6 Relating dynamic programming to linear programming

We continue here the discussion of 2.9 centred around the recursion given there. We assume that each x_i and each s_i can only take on a finite number of different values. Given the recursion of 2.9 we construct a *network* with *nodes* (i, s_i) and *arcs* x_i. An arc x_i leads *from* the node (i, s_i) to the node $(i-1, s_{i-1} = g_i(s_i, x_i))$. For a given i, there may be many nodes (i, s_i) because s_i can take on many values, but there is only one node $(0, s_0)$ and only one node (N, s_N). Let us call (N, s_N) the *source* and $(0, s_0)$ the *sink*. The length of the arc x_i between

(i, s_i) and $(i-1, s_{i-1})$ is given by $a_i(s_i, x_i)$. Thus we have constructed a *capacitated oriented* network.

We can now interpret $f_i(s_i)$ as the shortest distance from (i, s_i) to the sink. In particular $f_N(s_N)$, the value we are primarily interested in, is the shortest distance from source to sink.

The next step is to number the nodes arbitrarily, except that the sink gets the number 0 and the source gets the highest number, say n. Let c_{jk} be the length of the arc between node j and node k, so that if j corresponds to (i, s_i) and k to $(i-1, s_{i-1} = g_i(s_i, x_i))$ then $c_{jk} = a_i(s_i, x_i)$. For each arc introduce a variable z_{jk}. Then the shortest-distance problem for the source can be formulated as follows. Find

(7.6.1)

$$\min\left\{\Sigma c_{jk} z_{jk} : \Sigma z_{mk} - \Sigma z_{jm} = \begin{cases} 1 & \text{if } m = n, \\ 0 & \text{if } 1 \leqslant m \leqslant n-1;\ z_{jk} \geqslant 0 \\ -1 & \text{if } m = 0 \end{cases}\right\}$$

where the summations over j and k, m and k, or j and m are such that an arc exists from j to k, from m to k, or from j to m, respectively. The equivalence between the two formulations should be clear if we add the constraint that z_{jk} must be 0 or 1. For then (7.6.1) can be seen as the problem of finding the minimum distance from source to sink for a ball that is forced to enter the network at the source (see constraint with $m = n$) and after rolling through the network is forced to leave it at the sink (see constraint with $m = 0$). We claim, however, (without giving the not very difficult proof) that the condition that z_{jk} must be an integer may be ignored.

Clearly (7.6.1) is (finite-dimensional) linear programming, and its dual is the problem of finding

(7.6.2) $\max\{y_n^* - y_0^* : y_j^* - y_k^* \leqslant c_{jk}$ for all j, k such that there is an arc leading from node j to node $k\}$

(we leave the verification of this to the reader). Obviously the y_j^* are determined up to a constant. So let us take $y_0^* = 0$. With this restriction it is now easily verified that at optimum the y_j^* are nothing other than the $f_i(s_i)$ for appropriate (i, s_i). In particular, $y_n^* = f_N(s_N)$.

7.7 Separation by optimization

We have seen in the preceding chapters how important separation arguments are for the development of the theory of duality of optimization problems. In this very last section we go the other way around and try to derive a (weak) separation theorem from duality.

Separating two sets A and B can be reduced to separating $\{0\}$ from the set $A-B=\{x: x=a-b, a\in A, b\in B\}$, so let us assume that we are given a convex set of a locally convex topological vector space X such that $\operatorname{ri} C \neq \emptyset$, and $0\notin C$. We want to show that $x^*x \geqslant 0$ for some $x^* \neq 0$ and all $x\in C$.

Select a fixed $c_0 \in C$ and define f by $f(x)=\gamma$ if $x=\gamma c_0$, $\gamma\in R$, and $f(x)=+\infty$ otherwise. Then $\alpha=\inf\{f(x): x\in C\}\geqslant 0$, for otherwise for some $\gamma<0$ we must have that $\gamma c_0\in C$, but then

$$0=(1/(1-\gamma))\gamma c_0+(1-1/(1-\gamma))c_0\in C.$$

By Fenchel duality it follows that

$$\alpha=\inf\{f(x)-x^*x: x\in X\}+\inf\{x^*x: x\in C\}$$

for some x^*. Since $f(0)=0$ we have that $0\leqslant\alpha\leqslant\inf\{x^*x: C\}$, that is $x^*x\geqslant 0$ for all $x\in C$. Moreover $x^*\neq 0$, for otherwise $\alpha=\inf\{f(x): x\in X\}=-\infty$.

This shows that weak separation can be derived from duality, and since strong separation can be reduced to weak separation (see Appendix B) the same holds for the former. The final conclusion then is that in a sense we have not added much new in this book, except for presenting the idea of separating sets in entirely different formulations!

Appendix A

A.1 **Lemma.** Let $z_0^*, z_1^*, \ldots, z_n^*$ be linear functionals on a linear space Z, then either z_0^* is a linear combination of $z_1^*, \ldots, z_n^*$ or for some $z_0 \in Z$, $z_0^* z_0 = 1$, $z_1^* z_0 = \ldots = z_n^* z_0 = 0$.

Proof. The proof is by induction on n. The lemma is trivially true if $n = 0$. Assume it has been proved for $n = m-1$ and now let $n = m$. If e.g. z_1^* is a linear combination of $z_2^*, \ldots, z_m^*$, then by the induction assumption either z_0^* is a linear combination of $z_2^*, \ldots, z_m^*$ and hence of $z_1^* \ldots, z_m^*$, or for some z_0, $z_0^* z_0 = 1$, $z_2^* z_0 = \ldots = z_m^* z_0 = 0$, but then $z_1^* z_0$ is also zero. We may, therefore, assume that $z_1^*, \ldots, z_m^*$ are independent. Again by the induction assumption it follows from this that for all $i \geqslant 1$ there exists a z_i such that $z_i^* z_i = 1$ and $z_j^* z_i = 0$, $i, j = 1, \ldots, m$, $i \neq j$. Clearly for all $z \in Z$ we have that

$$z_i^* \left[z - \sum_{j=1}^{m} (z_j^* z) z_j \right] = 0.$$

When $z_0^*[z - \Sigma(z_j^* z) z_j] \neq 0$ for one such z we have finished. And if

$$z_0^*[z - \Sigma(z_j^* z) z_j] = z_0^* z - \Sigma z_j^* (z_0^* z_j) z = 0$$

for all $z \in Z$ then $z_0^* = \Sigma(z_0^* z_j) z_j$.

Theorem 3.2.6. If (z, z^*) is a dual pair then Z^* consists precisely of all σ-continuous linear functionals on Z.

Proof. Let z_0^* be any linear σ-continuous functional on Z. Then $|z_0^* z| \leqslant \alpha$ for some $\alpha < 1$, some finite subset $A^* = \{z_1^*, \ldots, z_n^*\}$ of Z^* and all z in a neighbourhood $U(A^*)$ of the origin. By the lemma either z_0^* is a linear combination of $z_1^*, \ldots, z_n^*$ and as such is an element of Z^*, or for some $z_0 \in Z$, $z_0^* z_0 = 1$ and $z_i^* z_0 = 0$, $i = 1, \ldots, n$, so that $z_0 \in U(A^*)$. But $|z_0^* z_0| = 1 > \alpha$, so that $z_0 \notin U(A^*)$, and only the first possibility remains, that is $z_0^* \in Z^*$.

Notice that Lemma A.1 has nothing to do with topology.

Appendix B

The first three of the following lemmas are concerned only with linear spaces and have, therefore, nothing to do with topology.

B.1 **Lemma** (Hahn–Banach). Let Z be a linear space, D a linear subspace of Z, F_0 a linear functional on D, $p: Z \to R$ a real function satisfying

(B.2a) $$p(z_1+z_2) \leqslant p(z_1)+p(z_2) \quad \text{if} \quad z_1, z_2 \in Z$$

(B.2b) $$p(\alpha z) = \alpha p(z) \quad \text{if} \quad z \in Z \quad \text{and} \quad \alpha \geqslant 0$$

(B.2c) $$F_0 z \leqslant p(z) \quad \text{if} \quad z \in D.$$

Then there exists a linear functional F on Z such that

(B.3) $$Fz = F_0 z \quad \text{if} \quad z \in D, \quad \text{and} \quad Fz \leqslant p(z) \quad \text{if} \quad z \in Z;$$

that is, such that F is an *extension* of F_0 to Z.

Proof. By Zorn's lemma and contradiction. Let F' be the maximal extension of F_0 such that $F'z = F_0 z$ if $z \in D$ and $F'z \leqslant p(z)$ if $z \in Z'$ for some subspace $Z' \supset D$, which means that there is no linear functional F'' on a subspace Z'' containing Z' strictly and such that $F''z = F_0 z$ if $z \in D$ and $F''z \leqslant p(z)$ if $z \in Z''$. Then $Z' = Z$; for if not then $z' \notin Z'$ for some $z' \in Z$. Given any x and any y in Z' we have

$$F'x - F'y = F'(x-y) \leqslant p(x-y) \leqslant p(x-z') + p(-y-z'),$$

and hence

$$\sup_{y \in Z'}\{-p(-y-z') - F'y\} \leqslant \inf_{x \in Z'}\{p(x+z') - F'x\},$$

so that for some real number c

(B.4) $$-p(-y-z') - F'y \leqslant c \leqslant p(x+z') - F'x \quad \text{if} \quad x, y \in Z'.$$

Define the linear functional F'' on the subspace spanned by Z' and z' by $F''(z+\alpha z') = F'z + \alpha c$, $z \in Z'$, $\alpha \in R$. Then $F''z = F'z = F_0 z$ if

$z \in D$ and $F''(z+\alpha z') \leqslant p(z+\alpha z')$ if $z \in Z'$, $\alpha \in R$. If $\alpha > 0$ the latter follows from (B.2*b*) and by taking $x = z/\alpha$ in (B.4), and similarly when $\alpha < 0$. But this means that F' is not maximal.

B.5 **Definition.** An element q of a set V of a linear space Z is an *internal point* of V if for all $z \in Z$ there exists an $\epsilon > 0$ such that $q+\delta z \in V$ for all δ with $|\delta| \leqslant \epsilon$.

B.6 **Lemma.** If V is a convex set of a linear space Z and 0 is an internal point of V, then there exists a nonzero linear functional on Z separating a given point $z_0 \notin V$ from V weakly in the noncontinuous sense (that is, not requiring that the functional must be continuous, which is a meaningless notion here).

Proof. We construct the so-called *Minkowski-functional* p (also termed *support function* which should not be confused with the conjugate of the indicator function) by

(B.7) $$p(z) = \inf\{\alpha : \alpha > 0, z/\alpha \in V\}, \quad z \in Z.$$

Then

(*a*) $p(z) \geqslant 0$ if $z \in Z$,

(*b*) $p(z) < +\infty$, since 0 is an internal point of V,

(*c*) $p(z_1+z_2) \leqslant p(z_1)+p(z_2)$ if $z_1, z_2 \in V$, for if $z_1/\alpha \in V$ and $z_2/\beta \in V$ then

$$(z_1+z_2)/(\alpha+\beta) = \alpha/(\alpha+\beta)(z_1/\alpha)+\beta/(\alpha+\beta)(z_2/\beta) \in V,$$

(*d*) $p(\alpha z) = \alpha p(z)$ if $z \in Z$ and $\alpha \geqslant 0$ (hence p is a *seminorm*),

(*e*) $p(z) \leqslant 1$ if $z \in V$, and

(*f*) $p(z_0) \geqslant 1$, because $z_0 \notin V$.

Further, let F_0 be defined by $F_0(\alpha z_0) = \alpha p(z_0)$ if $\alpha \in R$ and let $D = \{z : z = \alpha z_0, \alpha \in R\}$. Then by Lemma B.1 there exists a linear functional F on Z such that $F(\alpha z_0) = \alpha p(z_0)$ if $\alpha \in R$ and $Fz \leqslant p(z)$ if $z \in Z$. It follows that if $z \in V$ then $Fz \leqslant p(z) \leqslant 1$, but $Fz_0 = p(z_0) \geqslant 1$; hence F is a nonzero linear functional separating V and z_0 weakly in the noncontinuous sense.

B.8 **Lemma.** Let V and W be two nonempty convex sets of a linear space Z, such that $V \cap W = \emptyset$ and such that V has an internal point (which may be assumed to be 0). Then V and W can be separated weakly in the noncontinuous sense.

Proof. Take $z_0 \in W$ arbitrarily. Then 0 is an internal point of $V - W + \{z_0\}$ and this set is convex and does not contain z_0. By Lemma B.6 there exists a linear functional F on Z such that $F \neq 0$, $Fz \leqslant Fz_0$ if $z \in V - W + \{z_0\}$, i.e. such that $Fv \leqslant Fw$ if $v \in V$, $w \in W$.

Now we introduce topological concepts.

B.9 **Lemma.** An interior point of a set V of a locally convex topological vector space is an internal point of V.

Proof. Let q be an interior point of V. Then for some neighbourhood U of the origin of Z, $q + z \in V$ if $z \in U$. Given $z \in Z$ arbitrarily, there is an $\epsilon > 0$ such that $\epsilon z \in U$, so that $q + \delta z \in V$ if $|\delta| \leqslant \epsilon$.

B.10 **Lemma.** A linear functional F separating the sets V and W of a locally convex topological vector space is continuous if $\operatorname{int} V \neq \varnothing$.

Proof. We may assume that $0 \in \operatorname{int} V$. Take $z_0 \in W$. Then $0 \in \operatorname{int}(V - W + \{z_0\})$ and $F(v - w + z_0) \leqslant Fz_0$ if $v \in V$ and $w \in W$. But $0 \in V - W + \{z_0\}$ so that $Fz_0 \geqslant 0$ and $0 \in \operatorname{int}(V - W + \{z_0\})$, hence, for some $\epsilon > 0$, $z \in V - W + \{z_0\}$ if $|z| < \epsilon$, and for such z we have that $Fz \leqslant Fz_0$; but also $-Fz = F(-z) \leqslant Fz_0$, so that $|Fz| \leqslant Fz_0$. It follows that F is bounded, and hence continuous.

Proof of Theorem 3.3.3. Apply Lemmas B.8 and B.10 with V replaced by int V.

Proof of Theorem 3.3.2. Let V be compact. As V and W are closed and V is compact, $V - W$ is closed. But $0 \in V - W$, so $U \cap (V - W) = \varnothing$ for some neighbourhood U of the origin. Hence by Theorem 3.3.3, $Fu \leqslant F(v - w)$ if $u \in U$, $v \in V$, $w \in W$ for some nonzero linear functional F. Because $F \neq 0$, $Fu = \epsilon > 0$ for some $u \in U$ and it follows that $Fv \geqslant Fw + \epsilon$ for all $v \in V$ and all $w \in W$.

Next we turn to Theorems 3.3.5 and 3.3.6 involving *relative* interiors. First we quote the open mapping theorem. For its proof, which requires *Baire's category theorem*, we refer the reader to the literature.

B.11 **Lemma** (open mapping theorem). A linear continuous function from a Banach space *onto* another Banach space maps open sets onto open sets.

Proof of Theorem 3.3.6. Define $L(V)$ as in Definition 3.3.4 and define $L(W)$ similarly. Clearly $L(V)\times L(W)$ and $L(V)+L(W)$ are also Banach spaces and the linear continuous function s defined by $s(v, w) = v-w$, $v\in L(V)$, $w\in L(W)$, maps $L(V)\times L(W)$ onto $L(V)+L(W)$. By assumption $s(-v+\text{ri } V, -w+\text{ri } W)$ is not empty and by the open mapping theorem this set is open in $L(V)+L(W)$. Since $\text{ri } V \cap \text{ri } W = \emptyset$ we have that

$$\begin{aligned} -v+w \notin s(&-v+\text{ri } V, -w+\text{ri } W) \\ &= -v+w+\text{ri } V-\text{ri } W \subset L(V)+L(W). \end{aligned}$$

In parts (A) and (B) below we consider two cases.

(A) Assume that $-v+w\notin L(V)+L(W)$ for some $v\in V$ and some $w\in W$; then in Z we can separate $-v+w$ strongly from $L(V)+L(W)$, the latter being a closed set, and this means that V and W, too, can be separated in Z.

(B) Assume now that $-v+w\in L(V)+L(W)$ for some $v\in V$ and some $w\in W$; then in $L(V)+L(W)$ we can separate $-v+w$ weakly from $-v+w+\text{ri } V-\text{ri } W$, as follows from Theorem 3.3.3. The relevant linear continuous functional can now be extended to an element of Z^*. The latter follows from the Hahn–Banach theorem (Lemma B.1) by taking $D = L(V)+L(W)$ and $p(z) = 0$ for all $z\in Z$.

Proof of Theorem 3.3.5. As we do not actually need this theorem we take a number of facts for granted. Let k be the *canonical mapping* from Z onto the *quotient space* Z/M of Z and M, that is $k(z) = z+M$. Clearly k is linear and continuous, and since $L(V)+M = Z$ we have that $k(L(V)) = Z/M$. Both $L(V)$ and Z/M are Banach spaces on their own and $\text{ri } V$ is open in $L(V)$, so by the open mapping theorem $k(\text{ri } V)$ is open. Moreover $k(\text{ri } V) \cap k(W) = \emptyset$ since $(\text{ri } V-W) \cap M = \emptyset$ and everything is convex. Hence Theorem 3.3.3 implies the existence of a $q^*\in(Z/M)^*$ such that $q^* \neq 0$ and such that $q^*k(v) \leqslant q^*k(w)$ if $v\in\text{ri } V$ and $w\in W$. Putting $z = q^*k$ we have the required functional.

Appendix C

C.1 **Lemma.** A convex function f from a convex set C of a finite-dimensional space Z to R is continuous in $\operatorname{int} C$.

Proof. Let $z_0 \in \operatorname{int} C$ and let Z have dimension n. Then there exists a simplex S with vertices $z_1, \ldots, z_{n+1} \in C$ such that $z_0 \in \operatorname{int} S$. Let $\epsilon > 0$ be given, take λ such that $0 < \lambda < 1/(n+1)$ and

$$(n+1)\lambda \left\{ \sum_{i=1}^{n+1} |f(z_i)| + |f(z_0)| \right\} < \epsilon$$

and put $z_i' = (n+1)\lambda z_i + \mu z_0$ where $\mu = 1-(n+1)\lambda$, $i = 1, \ldots, n+1$.

The simplex S′ spanned by the z_i' is similar to S and oriented the same way because $z_i' - z_0 = (n+1)\lambda(z_i - z_0)$. There exists a $\delta > 0$ such that $z \in S'$ whenever $|z - z_0| < \delta$. Take such a z; then $z = \alpha_1 z_1' + \ldots + \alpha_{n+1} z_{n+1}'$ for certain $\alpha_i \geqslant 0$, $\Sigma \alpha_i = 1$.

Now

$$f(z) = f((n+1)\lambda(\Sigma \alpha_i z_i) + \mu z_0) \leqslant (n+1)\lambda \Sigma \alpha_i f(z_i) + \mu f(z_0),$$

since f is convex; hence

$$\text{(C.2)} \qquad f(z) - f(z_0) \leqslant (n+1)\lambda \Sigma\{|f(z_i)| + |f(z_0)|\} < \epsilon.$$

But $z' = 2z_0 - z$ is also an element of S', because $z' - z_0 = z_0 - z$ which implies that $|z' - z_0| = |z_0 - z| < \delta$. Hence similar to (C.2) we have the result

$$\text{(C.3)} \qquad f(z') - f(z_0) < \epsilon.$$

Since $z_0 = \frac{1}{2}(z + z')$ we further have that $f(z_0) \leqslant \frac{1}{2} f(z) + \frac{1}{2} f(z')$ so that $f(z) - f(z_0) \geqslant -f(z') + f(z_0)$ and by (C.3) this is greater than $-\epsilon$. Combining this with (C.2) we find that $|f(z) - f(z_0)| < \epsilon$.

Appendix D

D.1 **Lemma.** Let $X = R_n$ with the nonnegative orthant as the positive cone, Y be a locally convex topological vector space, and $Z: X \to Y$ a linear continuous mapping. Then

(D.2) $$G = \{y: y = Ax, x \geqslant 0, x \in X, y \in Y\}$$

is closed.

Proof. By induction on n.

First let $n = 1$ and assume that $y^i = Ax^i$ for some $x_i \geqslant 0$ converges to y. When $A = 0$ then $y = 0 = A0$ so that $y \in G$. When $A \neq 0$ but x^i converges to some x (for a subsequence) then $x \geqslant 0$ and $y = Ax$ so that again $y \in G$. In the remaining cases we have that $A \neq 0$ and that x^i converges to $+\infty$ (for a subsequence). This, however, is not possible since $A1 \neq 0$, so that $y^i = (x^i)A1$ does not converge.

Now let $n > 1$ and let y^i, y and x^i be as before. Put $x^i = (\xi_1^i, \ldots, \xi_n^i)$ and $a_k = A(0, \ldots 0, 1, 0, \ldots 0)$ with the 1 at position k, then $y^i = \Sigma_k \xi_k^i a_k$. Again when x^i converges to some x (for a subsequence) we have finished. If this is not the case then ξ_k^i tends to $+\infty$ for some k and some subsequence. Permuting the components of x, if necessary, we may assume that $k = n$ and moreover that $\xi_1^i \leqslant \xi_2^i \leqslant \ldots \leqslant \xi_n^i$ and that ξ_k^i/ξ_n^i converges for all k (for a subsequence). Define μ_k by

$$\mu_k = \lim_{i \to \infty} \xi_k^i/\xi_n^i, \quad k = 1, \ldots, n.$$

Then $y^i/\xi_n^i = \Sigma_k \xi_k^i \alpha_k/\xi_n^i$ implies that

(D3) $$\Sigma_k \mu_k \alpha_k = 0;$$

in other words, the rows of A are dependent.

For each i let q be such that $\mu_q > 0$ and

$$\xi_q^i/\mu_q = \min_{\mu_k > 0} \xi_k^i/\mu_k.$$

Such a q exists because $\mu_n = 1$, but it depends on i. Since X is finite-dimensional, however, there exists a subsequence such that q does not depend on i. For simplicity suppose that $q = 1$ for all i. Then $\xi_k^i - \xi_1^i \mu_k/\mu_1 \geqslant 0$ if $\mu_k > 0$, but this inequality also holds if $\mu_k \leqslant 0$, simply because $\xi_k^i \geqslant 0$ and $\mu_1 > 0$. By means of (D.3) we now eliminate a_1 in $y^i = \Sigma_k \xi_k^i a_k$ and we obtain

(D.4) $$y_i = \Sigma_{k\geqslant 2}(\xi_k^i - \xi_1^i \mu_k/\mu_1)\, a_k = \Sigma_{k\geqslant 2}\zeta_k^i a_k \quad \text{with} \quad \zeta_k^i \geqslant 0,$$

which means that we have reduced the problem to the case where $X = R_{n-1}$.

D.5 **Corollary.** If D is a linear subspace of R_n and Q is the nonnegative orthant of R_n, then $G = Q + D$ is closed.

Proof. D can be written parametrically as $D = \{z: z = Bv \text{ for some } v\}$ where B is an appropriate linear mapping. Replace v by $v_1 - v_2$ with $v_1 \geqslant 0$ and $v_2 \geqslant 0$; then if we denote the identity by I we have that $G = \{y: y = (I, B, -B)(x, v_1, v_2)^t \text{ for some } (x, v_1, v_2) \geqslant 0\}$.

Appendix E

Before giving the proof of Lemma 5.1.7 we state two other lemmas.

E.1 **Lemma** (Banach inverse theorem). The inverse of a linear continuous one-to-one mapping A from a Banach space X' onto another Banach space $\tilde{X}$ is linear and continuous.

Proof. By the open mapping theorem $(A^{-1})^{-1} = A$ maps open sets onto open sets, therefore A^{-1} is continuous. And the linearity of A^{-1} is also easily shown.

E.2 **Lemma** (generalized mean value theorem). If s is a Fréchet-differentiable in a neighbourhood of x_o then for h close enough to zero

(E.3) $$|s(x_o+h)-s(x_o)| \leqslant |h| \sup_{0<\theta<1} |s'(x_o+\theta h)|.$$

The latter lemma is a reformulation of the first part of Lemma 5.1.6(*d*).

Proof of Lemma 5.1.7.

(A) Let N be the *nullspace* of $q'(x_o)$, that is

(E.4) $$N = \{x\colon q'(x_o)x = 0\}$$

and consider the Banach space $X' = X/N$ with elements the sets $x' = x+N$ and with norm $|x'| = \inf_{n\in N}|x+n|$. Clearly x may be replaced here by any other element of $x+N$. Secondly define the linear continuous mapping $A\colon X' \to \tilde{X}$ by

(E.5) $$Ax' = q'(x_o)x \quad \text{with} \quad x \in x'$$

(again the choice of x in x' does not matter). Since $q'(x_o)$ is onto, so is A. Moreover A^{-1} exists, because given x there is only one set

x' such that $x \in x'$. Hence by Lemma E.1, A^{-1} is linear and continuous. Thirdly choose $\delta > 0$ such that

(E.6) $|q'(x) - q'(x_o)| \leqslant \epsilon = 1/4|A^{-1}|$ whenever $|x - x_o| \leqslant \delta$,

which is possible because q' is continuous at x_o.

(B) In what follows we construct sequences $a_i \in X$ and $a'_i \in X' = X/N$ such that $a'_i = a'_{i-1} - A^{-1}q(x_o + \lambda x_1 + a_{i-1})$ and such that a_i converges to some a, and a'_i to some a'. We will be able to select these sequences such that $a = o(\lambda)$, so that $a' = a' - A^{-1}q(x_o + \lambda x_1 + o(\lambda))$ as follows from the continuity of q in a neighbourhood of x_o and that of A^{-1}. So

$$A^{-1}q(x_o + \lambda x_1 + o(\lambda)) = 0;$$

hence $q(x_o + \lambda x_1 + o(\lambda)) = 0$.

(C) Let $\lambda \in R$, $a_i \in X$, $a'_i \in X'$ such that $a_0 = 0$, $a'_0 = N$ and

(E.7) $$a'_i = a'_{i-1} - A^{-1}q(x_o + \lambda x_1 + a_{i-1}),$$

(E.8) $$a_i \in a'_i,$$

(E.9) $$|a_i - a_{i-1}| \leqslant 2|a'_i - a'_{i-1}|,$$

for $i = 1, 2, \ldots$ The choice of a_i such that (E.8) and (E.9) hold is possible since by the definition of norm in $X' = X/N$ we have $|a'_i - a'_{i-1}| = \inf_{n \in N}|x + n|$ with $x \in a'_i - a'_{i-1}$. As $a_0 \in a'_0$ we may assume $a_{i-1} \in a'_{i-1}$ and take $x = y - a_{i-1}$ with $y \in a'_i$. Then

$$|a'_i - a'_{i-1}| = \inf_{n \in N}|y + n - a_{i-1}| = \inf_{z \in a'_i}|z - a_{i-1}|,$$

so that indeed we can select $a_i \in a'_i$ such that (E.9) holds.

In particular, consider a_1. We have that

$$|a_1| \leqslant 2|a'_1 - a'_0| \leqslant 2|A^{-1}| \cdot |q(x_o + \lambda x_1)|.$$

But

$$q(x_o + \lambda x_1) = q(x_o) + q'(x_o)\lambda x_1 + o(\lambda) = o(\lambda)$$

since by assumption $q(x_o) = 0$ and $q'(x_o)x_1 = 0$. Therefore

(E.10) $|a_1| \leqslant 2|A^{-1}| \cdot |o(\lambda)| \leqslant \frac{1}{4}\delta$ if λ is small enough.

This we shall use below.

(D) Because of (E.5) it follows that

$$Aa'_{i-1} = q'(x_o)a_{i-1} \quad \text{since} \quad a_{i-1} \in a'_{i-1},$$

so (E.7) leads to

$$a_i' = -A^{-1}[q(x_0+\lambda x_1+a_{i-1})-q'(x_0)a_{i-1}]$$

and hence to

$$\text{(E.11)}\quad \begin{aligned} a_i'-a_{i-1}' = -A^{-1}[&q(x_0+\lambda x_1+a_{i-1}) \\ &-q'(x_0)(x_0+\lambda x_1+a_{i-1})-q(x_0+\lambda x_1+a_{i-2}) \\ &+q'(x_0)(x_0+\lambda x_1+a_{i-2})]. \end{aligned}$$

To (E.11) we apply Lemma (E.2) with $h = a_{i-1}-a_{i-2}$ and

$$s(x) = -A^{-1}[q(x+\lambda x_1+a_{i-2})-q'(x_0)(x+\lambda x_1+a_{i-2})],$$

so that $$s'(x) = -A^{-1}[q'(x+\lambda x_1+a_{i-2})-q'(x_0)].$$

The result is

$$\text{(E.12)}\quad \begin{aligned} |a_i'-a_{i-1}'| \leqslant{} & |a_{i-1}-a_{i-2}| \\ & \times \sup_{0<\theta<1}|-A^{-1}[q'(x_0+\lambda x_1+a(\theta))-q'(x_0)]|, \end{aligned}$$

where $a(\theta) = a_{i-2}+\theta(a_{i-1}-a_{i-2})$. Using the continuity of q' at x_0 we will show that (E.12) implies that $|a_i| \leqslant \frac{1}{2}\delta$ for all i. Clearly $|a_0| \leqslant \frac{1}{2}\delta$ and by (E.10) $|a_1| \leqslant \frac{1}{2}\delta$. So let us assume that $|a_0| \leqslant \frac{1}{2}\delta, \ldots, |a_{i-1}| \leqslant \frac{1}{2}\delta$. Then also $|a(\theta)| \leqslant \frac{1}{2}\delta$. Taking λ so small that $|\lambda x_1| \leqslant \frac{1}{2}\delta$ we see that the norm of the expression in square brackets in (E.12) is smaller than ϵ, defined in (E.6). Hence

$$\begin{aligned} |a_i'-a_{i-1}'| &\leqslant |a_{i-1}-a_{i-2}|\cdot|A^{-1}|\epsilon \\ &\leqslant 2|a_{i-1}'-a_{i-2}'|\cdot|A^{-1}|\epsilon \leqslant \tfrac{1}{2}|a_{i-1}'-a_{i-2}'| \end{aligned}$$

and

$$\begin{aligned} |a_i-a_{i-1}| &\leqslant 2|a_{i-1}'-a_{i-2}'| \\ &\leqslant 2|a_{i-1}-a_{i-2}|\cdot|A^{-1}|\epsilon \leqslant \tfrac{1}{2}|a_{i-1}-a_{i-2}|, \end{aligned}$$

that is

$$\text{(E.13)}\qquad |a_i-a_{i-1}| \leqslant \tfrac{1}{2}|a_{i-1}-a_{i-2}|.$$

Consequently $|a_i-a_{i-1}| \leqslant (1/2^{i-1})\,|a_1|$ and

$$|a_i| \leqslant |a_1|+|a_2-a_1|+\ldots+|a_i-a_{i-1}| \leqslant 2|a_1| \leqslant \tfrac{1}{2}\delta$$

as follows from (E.10). Hence we have indeed that $|a_i| \leqslant \frac{1}{2}\delta$ for all i. But this means that (E.13) holds for all i, from which we see that a_i

must converge to some $a \in X$, since X is *complete*. This a satisfies the condition that

$$|a| \leqslant 2|a_1| \leqslant 4|A^{-1}| \cdot |o(\lambda)|,$$

again by (E.10). Finally (E.13) also holds if we put primes everywhere, so that a_i' converges to some $a' \in X'$; hence from (E.7) and the continuity of q and A^{-1} we find that $q(x_0 + \lambda x_1 + a) = 0$, where $a = o(\lambda)$.

Notice that $o(\lambda)$ sometimes denotes an element of $\tilde{X}$ (e.g. in part (C)) and sometimes an element of X (e.g. in part (B)).

Comments on the text and related literature

The material of the book is based on a number of topological and functional analytical results, which can be found in e.g. Berge (1963), Dunford & Schwartz (1957), Liusternik & Sobolev (1961), Robertson & Robertson (1973). Other facts of a general nature are provided by Eggleston (1958), Ekeland & Temam (1976), Rockafellar (1970*b*) (all three about convexity), and Smart (1974) (about fixed point theory).

Books or (series of) papers containing a broad account of the theory of optimization (which some people may prefer to call the mathematics of *operations research*) are those by Balakrishnan (1971), Balinski & Wolfe (1975), Barbu & Precupanu (1978), Blum & Oettli (1975), Canon, Cullum & Polak (1970), Collatz & Wetterling (1966), Craven (1978), Dempster (1975), Dubovitskii & Milyutin (1965), Duffin, Peterson & Zener (1967), Geoffrion (1971), Girsanov (1972), Gol'šteĭn (1972), Göpfert (1973), Hadley (1964), Halkin (1974), Hestenes (1966), Holmes (1972), Ioffe & Tikhomirov (1968), Kall (1976), Luenberger (1969, 1973), Mangasarian (1969), Neustadt (1974), Pallu de la Barrière (1967), Pontryagin, Boltyanskii, Gramkelidze & Mishchenko (1962), Rockafellar (1970*b*, 1974*b*), Ritter (1969, 1970), Russell (1970), Stoer & Witzgall (1970), Tabak & Kuo (1971), Vajda (1961), Varaiya (1967), Wagner (1969), Wets (1976).

As the book deals with only a limited number of approaches to the theory of optimization, we refer the reader to a few other attempts to set up such a theory: Balinkski & Wolfe (1975), Bazaraa & Goode (1972), Ben-Tal, Ben-Israel & Zlobec (1976), Clarke (1976), Das (1975), Duffin (1973), Ekeland (1979), Elster & Nehse (1974, 1975, 1976) (these authors follow a purely *algebraic* approach, not being concerned about the boundedness of the desired linear functionals), Girsanov (1972), Halkin (1974), Hestenes (1966), Hoàng Tuy

(1974*a*), Holmes (1972), Jongen (1977), Levine & Pomerol (1974), Mangasarian & Ponstein (1965), Neustadt (1974), Nieuwenhuis (1978), Peterson (1973), Rockafellar & Wets (1976*a*, *b*), Ritter (1967, 1969, 1970), Roode (1968), Smith & Vandelinde (1972).

Chapter 1. Example 1.2.4, about the polygonal steel plate, has been taken from Collatz & Wetterling (1966).

Production planning problems such as Example 1.2.5 occur in economics in a wide range of sizes and degrees of complexity. Methods of solving these problems have been improved to such an extent that problems with thousands of constraints and variables can be solved within a reasonable time and with reasonable accuracy.

Example 1.2.6 is just one simple example of a large field of scientific interest: *optimal control.* This discipline is older than that of mathematical programming (which is the basic point of view in the book), and its original tools – notably the calculus of variations – are quite different from those used in mathematical programming. This makes it all the more interesting to observe the attempts to treat optimal control from the standpoint of mathematical programming; see e.g. Canon, Cullum & Polak (1970), Ekeland & Temam (1976), Girsanov (1972), Halkin (1974), Ioffe & Tikhomirov (1968), Luenberger (1969), Neustadt (1974), Ponstein (1968), Rockafellar (1970*a*, 1974*b*), Tabak & Kuo (1971), Van Slyke & Wets (1968), and section 7.3. Other references of interest here are Hestenes (1966) and Pallu de la Barrière (1967).

Example 1.2.7, too, is just an elementary example of a rapidly growing field, that of *game theory*. An offshoot of this is *differential game theory*, which is the game theoretic counterpart of optimal control.

Stochastic optimization problems such as the one of Example 1.2.8 were usually solved by means of dynamic programming. For a mathematical programming approach see Kall (1976), Rockafellar & Wets (1967*a*, *b*), and section 7.4.

Example 1.2.9 shows an application of optimization to *statistical decision theory*, for which we refer the reader to Schaafsma (1970), which contains some more references.

Chapter 2. The idea of basing the theory of optimization on the *perturbation* of a given problem can be found in several contributions to the literature (see e.g. Geoffrion 1971) but has been developed systematically by Rockafellar (1970*b*, 1974*b*). Two important branches that emanate from this are Lagrangian duality (by perturbing *equalities* and *inequalities*) and Fenchel duality (by perturbing *sets*). Among the remaining possibilities are those where one first transforms the problem and then applies one of these main forms. An example is *geometric programming* (see e.g. Duffin, Peterson & Zener 1967; Peterson 1973), where the problem is to find

$$\inf\{f(z): z = (\zeta_1, \ldots, \zeta_n) \in R_n, g_i(z) \leqslant 1, i = 1, \ldots, m, z > 0\},$$

Here f as well as all g_i are sums of products of (positive or negative) powers (integral or not) of the ζ_j and a positive coefficient. Notice the strict inequality in $z > 0$. This somewhat inconvenient type of constraint is removed by putting $\zeta_j = \exp \xi_j$. At the same time the logarithm is taken of f as well as of all g_i and suitable variables σ_{ij} are introduced. After this transformation the problem becomes that of finding

$$\inf\{\ln(\Sigma_k \exp \sigma_{ok}) : \ln(\Sigma_k \exp \sigma_{ik}) \leqslant 0,$$
$$\Sigma_j \alpha_{ikj} \xi_j - \sigma_{ik} + \delta_{ik} = 0,\ i = 1, \ldots, m,\ k = 1, 2, \ldots\}$$

with α_{ikj} and δ_{ik} known, and ξ_j and σ_{ik} variable. This is then treated by Lagrangian duality; that is, in the same way that problem 2.1.1 is treated.

The term '*bifuncton*' has been taken from Rockafellar (1970*b*), where the reader can also find the development of *subdifferentiability* and the abstract introduction of the *Lagrangian* as in Definition 2.5.1. More about subdifferentiability can be found in Bazaraa, Goode & Nashed (1974) and Taylor (1973).

Some basic results of *optimal control* by means of mathematical programming can be obtained by simply generalizing equations like (2.8.4) to equations valid in sufficiently general spaces. Although the special case of (2.8.6) is usually named after Kuhn and Tucker, it is only fair to mention an apparently overlooked (master!) thesis by Karush (1939!) which provides these conditions. For this reason some authors prefer to speak of the Karush–Kuhn–Tucker conditions.

An account of *dynamic programming*, introduced by Bellman, can be found in, for example, Hadley (1964) and Wagner (1969).

Chapter 3. This chapter is an amalgamation of ideas made known by Rockafellar (1970*b*) and Van Slyke & Wets (1968). The influence of the former author is through notions like bifunction and perturbation, whereas the usage of the sets T and V as defined in 3.5 is from the latter authors. The present author believes that these sets quite clearly illustrate the entire matter visually. Moreover the main theorem (3.6.1) adapts itself to many practical problems giving a necessary and sufficient condition except for cases which in practice will not occur (i.e. 'infeasible' problems). Notice, however, our comments regarding chapter 4. For material similar to that of this chapter see also Varaiya (1967).

Apart from *convexity* (and strict convexity introduced in 6.1) various weaker forms of convexity are known, a review of which has been given by Ponstein (1967). The most important of these is probably *quasi-convexity* (see again 6.1), because its definition does not even require the linearity of the space involved, far less the differentiability. Another one is *pseudo-convexity*. A function h is pseudo-convex if $(x_2 - x_1)\nabla h(x_1) \geqslant 0$ implies that $h(x_2) \geqslant h(x_1)$. Careful examination of what is needed sometimes reveals that we can do with less than convexity. In some locally sufficient optimality conditions (chapter 5) it is possible to replace the convexity of the objective function by pseudo-convexity and that of the constraint functions by quasi-convexity.

The basic material on *topology* can be found in, for example, Robertson & Robertson (1973). Whereas our *separation theorems* involve two sets, Dubovitskii & Milyutin (1965) consider a separation theorem involving $k+1$ sets $V_0, V_1, \ldots, V_k \subset X$, all nonempty and convex and with $0 \in \operatorname{cl} V_i$ for all i and V_i open for $i \geqslant 1$. Then $\bigcap_{i=0}^{k} V_i = \varnothing$ if and only if there exist $x_0^*, \ldots, x_k^*$ such that $(x_0^*, \ldots, x_k^*) \neq 0$, $x_0^* + \ldots + x_k^* = 0$ and $x_i^* x_i \geqslant 0$ whenever $x_i \in V_i$, $i \geqslant 0$. Notice that all 'multipliers' x_i^* operate on X, which reminds us of Theorem 3.14.30, where the so-called Dubovitskii–Milyutin 'formalism' might indeed be applicable.

The sufficiency conditions of 3.7 are straight generalizations of results by Van Slyke & Wets (1968). For other conditions, see Rockafellar (1974*b*), from which a result akin to Theorem 3.8.16 is also taken. This theorem involves the cone $K(\gamma)$ of Definition 3.8.5 which is from Klee (1948). The results of 3.11 to 3.15 are from sources mentioned earlier or are obvious generalizations of them. The methods of proof, however, might deviate here and there from the ones of these sources. When X and Y are Banach spaces (or at least *barrelled* spaces) conditions (3.11.11) to (3.11.13) can be weakened to: the sets C and $\{(x, y): g(x) \geqslant y\}$ are closed, and for all $y \in Y$ there exists an $\epsilon > 0$ and an $x \in C$ such that $g(x) \leqslant \epsilon y$ (see Rockafellar 1974*b*). If, in Theorem 3.13.2, $g(x) \leqslant 0$ consists of the combination of $g_1(x) \leqslant 0$ and $Ax \leqslant a$, then the regularity conditions can be weakened in that (3.13.5) becomes: $g_1(\hat{x}) < 0$, $A\hat{x} \leqslant a$, $B\hat{x} = b$ for some $\hat{x} \in \operatorname{ri} C$, if both $Ax \leqslant a$ and $Bx = b$ consist of a finite number of scalar constraints. Hence it is not then necessary to require that $A\hat{x} < a$. This is shown in Ponstein & Klein Haneveld (1975) and is the generalization of a similar result for R_n in Rockafellar (1970*b*). In the latter book the proof is given by means of a special separation theorem, whereas in the proof of the former paper the problem is first transformed into one to which the main theory can be applied, leading to a comparatively short proof. For an alternative proof of Theorem 3.13.8 see Nieuwenhuis (1978), from which Example 3.14.27 is taken. More about Fenchel duality can be found in McLinden (1974).

The contents of 3.16 are a continuation of the results by Magnanti (1974).

Chapter 4. The results of this chapter follow from those of chapter 3. This is not necessary, however, as can be seen from Rockafellar (1970*b*, 1974*b*), where the emphasis is strongly on the theory of *conjugate functions*, leading to a very *symmetric* theory. Although in essence equivalent to the theory of chapter 3 it has the important mathematical advantage of bearing a theory of *converse duality* along with that of 'straight' duality without any further ado.

The account in 4.6 on *variational inequalities* and the *complementarity problem* being rather compact, we mention here several papers on these subjects: Allen (1977), Eaves (1971), Karamardian (1969,

1971, 1972, 1976), Mancino & Stampaccia (1972), Mangasarian (1974, 1975*a*), Rockafellar (1976*b*), Saigal (1976).

Chapter 5. In this chapter we again come across the (Karush)–Kuhn–Tucker conditions which in generalized form become what is often called the *(pre)minimum principle*. The original (long) proof of this principle is given in Pontryagin *et al.* (1962). Later other ways of proving the principle were found. In Euclidean spaces it takes the form of (5.1.4).

A number of the results of this chapter can be seen as following from the approximation of the functions involved by linear ones. Since in the book *functions* come to the fore (as in Lagrangian duality) and then *sets* (as in Fenchel duality) it is only natural to see whether sets can be linearized as well. The usual tool for this is the *cone*; see e.g. Bazaraa & Goode (1972), Bazaraa, Goode & Nashed (1974) (where applications of the cone of tangents are considered), Hestenes (1966), Hoàng Tuy (1974*a*) (the cone of internal directions), Nieuwenhuis (1978).

A new basic tool is the *linearization lemma* 5.1.7 due to Liusternik & Sobolev (1961). The naming is from Nieuwenhuis (1978), who generalized the Liusternik & Sobolev result given in (5.1.8) by allowing this equation to be replaced by $q(x_0+\lambda x_1+o(\lambda))\in \tilde{P}$, with $\tilde{P}$ the positive cone of $\tilde{X}$. The proof of this generalization requires the introduction of *convex processes*, treated in e.g. Rockafellar (1970*b*). Other authors, instead of applying the linearization lemma, use an *inverse function theorem*. I believe with Hoàng Tuy and Nieuwenhuis that a linearization lemma nicely fits problems with (in)equality constraints. This is demonstrated by the procedure in Luenberger (1969), where an additional result is required before the inverse function theorem can be applied properly. Ironically enough, this author copies the Liusternik & Sobolev proof of their linearization lemma, except for a small but important detail somewhere at the beginning of the proof, and obtains an inverse functional result. Thus it would seem that he modifies the proof of the more appropriate theorem to get that of the less appropriate one.

As pointed out in the preface we have not dealt with *constraint qualifications* that are not only sufficient but also necessary for a certain class of objective functions, as did Gould & Tolle (1971) and Nieuwenhuis (1978). In Mangasarian & Ponstein (1965) a necessary and sufficient constraint qualification is given for just one problem, but its verification is not necessarily easy. Further, see Gould & Tolle (1975), Mangasarian (1969) (containing some seven constraint qualifications) and Slater (1951*a*).

Theorem 5.4.18 is applied to *optimal control* in *Hilbert spaces* in Ponstein (1968).

Results on *second-order conditions* can be found in Luenberger (1973) and McCormick (1967).

The *sufficiency theorems* of this chapter are related to converse duality (see e.g. Craven 1975; Mangasarian & Ponstein 1965).

Chapter 6. The relationship between saddle-points and *fixed points* as presented is well-known, as is that between *equilibrium points of games* and fixed points. In the case of two players the usual assumption is that x must be in $C \subset X$ and that z must be in $D \subset Z$. A more general case where $C \times D \subset X \times Z$ is replaced by a convex set of $X \times Z$ is solved in Ponstein (1966).

Theorem 6.1.9 is due to Kaulgud & Pai (1975) which contains further references to this sort of result. The proof of Brouwer's *fixed point theorem* (Lemma 6.1.15) follows the combinatorial argument given by Sperner and employed by Knaster, Kuratowski & Mazurkiewicz (1929). Although this proof is ingenious Kuhn (1960) ponders the question of whether it is also 'natural' and then tries to replace simplicial subdivisions (of a simplex) by cubical subdivisions (of a cube). Lemma 6.1.16 is by Kakutani (1941) and the final result, Theorem 6.1.18, has been shown by several authors (see e.g. Glicksberg 1952; Ky Fan 1952; and also Berge 1963). This theorem immediately leads to a *minimax theorem* (Theorem 6.1.22), but is of little use if one wants to prove that inf sup = sup inf when min and/or max do not exist. This is the subject of 6.2 and is based on Hoàng Tuy (1974*b*) (which strictly speaking is not concerned with minimax, as the title of the paper suggests, but with sup inf = inf sup, as is Sion (1958)). In fact Hoàng Tuy's result is more general than Theorem 6.2.2 for the simple reason that the condition of quasi-convex-concavity can be relaxed. In relaxed form the result includes Sion's result as well as some other results. The reason why we have assumed the quasi-convex-concavity is that this assumption adds much to the clarity of the theorem and its proof, without losing much of the content of the theorem and the point of the proof.

Modified (or *augmented*) *Lagrangians* is a subject with increasing practical interest. Our presentation is based on Arrow, Gould & Howe (1973). A large number of modifications are reviewed in Pierre & Lowe (1975). Other papers or books of interest in this respect are Buys (1972), Gould (1969), Mangasarian (1975*b*), Rockafellar (1974*a*, 1976*a*), Roode (1968).

Chapter 7. The comments here are a continuation of those made on chapter 1.

To treat inequalities such as the one in 7.1 by optimization theory is fairly obvious, in particular if one is after so-called '*sharp*' *inequalities* where equality occurs for certain values of the variables involved.

For the representation of the continuous linear functionals employed in 7.2 and 7.3 the reader may consult Dunford & Schwartz (1957)..

The book by Pallu de la Barrière (1967) is one of the relatively few discussing the difference between *open* and *closed loop control*.

Results about *theorems of the alternative* can be found in Bazaraa (1973), Craven & Mond (1973), Farkas (1902), Gordan (1873), McLinden (1975), Motzkin (1936), Slater (1951*b*), Stiemke (1915), Tucker (1956).

Linear programming and *dynamic programming* are among the many subjects of *operations research* (see e.g. Wagner 1969).

Appendixes. The contents of Appendix A can be found in many textbooks on topology; see e.g. Robertson & Robertson (1973). The separation theorems of Appendix B are from Dunford & Schwartz (1957) when interiors of sets are involved, and from Nieuwenhuis (1978) when we are concerned with the relative interiors of sets. Baire's category theorem, too, can be found in Dunford & Schwartz. Appendix E has been taken from Liusternik & Sobolev (1961).

References

This list is by no means exhaustive. In this respect much more information is supplied by Blum & Oettli (1975) and Stoer & Witzgall (1970). *JOTA* is the abbreviation of *Journal of Optimization Theory and Applications*, and *SIAM* that of *Society of Industrial and Applied Mathematics*.

Allen, G. (1977). Variational inequalities, complementarity problems, and duality theorems. *J. Math. Anal. Applics.* **58**, 1–10.

Arrow, K. J., Gould, F. J. & Howe, S. M. (1973). A general saddle-point result for constrained optimization. *Mathematical Programming* **5**, 225–34.

Balakrishnan, A. V. (1971). *Introduction to optimization theory in a Hilbert space.* Lecture Notes in Operational Research and Mathematical Systems 42. Springer, Berlin.

Balinski, M. L. & Wolfe, P. (eds.) (1975). *Nondifferential optimization.* Mathematical Programming Study 3.

Barbu, V. & Precupanu, Th. (1978). *Convexity and optimization in Banach spaces.* Sijthoff and Noordhoff, Alphen aan de Rijn.

Bazaraa, M. S. (1973). A theorem of the alternative with application to convex programming: optimality, duality, and stability. *J. Math. Anal. Applics.* **41**, 701–15.

Bazaraa, M. S. & Goode, J. J. (1972). Necessary optimality criteria in mathematical programming in the presence of differentiability. *J. Math. Anal. Applics.* **40**, 609–21.

Bazaraa, M. S., Goode, J. J. & Nashed, M. Z. (1974). On the cones of tangents with applications to mathematical programming. *JOTA* **13**, 389–426.

Ben-Tal, A., Ben-Israel, A. & Zlobec, S. (1976). Characterization of optimality in convex programming without a constraint qualification. *JOTA* **20**, 417–37.

Berge, C. (1963). *Topological spaces.* The MacMillan Company, New York.

Blum, E. & Oettli, W. (1975). *Mathematische Optimierung.* Springer, Berlin.

Buys, J. D. (1972). *Dual algorithms for constrained optimization problems* (thesis). University of Leiden.

Canon, M. D., Cullum, C. D. & Polak, E. (1970). *Theory of optimal control and mathematical programming.* McGraw-Hill, New York.

Clarke, F. H. (1976). A new approach to Lagrange multipliers, *Math. O.R.* **1**, 165–74.

Collatz, L. & Wetterling, W. (1966). *Optimierungsaufgaben.* Springer, Berlin.

Craven, B. D. (1975). Converse duality in Banach space. *JOTA* **17**, 229–38.

Craven, B. D. (1978). *Mathematical programming and control theory*. Chapman and Hall, London.
Craven, B. D. & Mond, B. (1973). Transposition theorems for cone-convex functions. *SIAM J. appl. Math.* **24**, 603–12.
Das, P. C. (1975). Constraint optimization problems in Banach space. *JOTA* **17**, 279–91.
Dempster, M. A. H. (1975). *Elements of optimization*. Chapman and Hall, London.
Dubovitskii, A. Y. & Milyutin, A. A. (1965). Extremum problems in the presence of restrictions. *USSR Comp. Math. math. Phys.* **5**, 1–80 (English translation).
Duffin, R. J. (1973). Convex analysis treated by linear programming. *Mathematical Programming* **4**, 125–43.
Duffin, R. J., Peterson, E. L. & Zener, C. (1967). *Geometric programming*. Wiley, New York.
Dunford, N. & Schwartz, J. T. (1957). *Linear operators*, part I. Interscience, New York.
Eaves, B. C. (1971). On the basic theorem of complementarity. *Mathematical Programming* **1**, 68–75.
Eggleston, H. G. (1958). *Convexity*. Cambridge University Press.
Ekeland, I. (1979). Nonconvex minimization problems. *Bull. Am. Math. Soc.* (new ser.) **1**, 443–74.
Ekeland, I. & Temam, R. (1976). *Convex analysis and variational problems*. North-Holland, Amsterdam (English translation from the French).
Elster, K.-H. & Nehse, R. (1974). Zum Dualitätssatz von Fenchel. *Math. Operationsforsch. Statist.* **5**, 269–80.
Elster, K.-H. & Nehse, R. (1975). Konjugierte Operatoren und Subdifferentiale. *Math. Operationsforsch. Statist.* **6**, 641–57.
Elster, K.-H. & Nehse, R. (1976). Zur Trennung konvexer Mengen mittels linearer Operatoren. *Math. Nachrichten* **71**, 171–81.
Farkas, J. (1902). Über die Theorie der einfachen Ungleichungen. *J. reine angewandte Math.* **124**, 1–24.
Geoffrion, A. M. (1971). Duality in nonlinear programming: a simplified applications-oriented development. *SIAM Rev.* **13**, 1–37.
Girsanov, I. V. (1972). *Lectures on mathematical theory of extremum problems*. Lecture Notes in Economics and Mathematical Systems 67. Springer, Berlin.
Glicksberg, I. L. (1952). A further generalization of the Kakutani fixed point theorem, with application to Nash equilibrium points. *Proc. Am. Math. Soc.* **3**, 170–4.
Gol'šteĭn, E. G. (1972). *Theory of convex programming*. Translations of mathematical monographs 36. Am. Math. Soc. (English translation from the Russian), Providence, R.I.
Göpfert, A. (1973). *Mathematische Optimierung in allgemeinen Vektorräumen*. Teubner, Leipzig.
Gordan, P. (1873). Über die Auflösungen linearer Gleichungen mit reellen Coefficienten. *Math. Ann.* **6**, 23–8.
Gould, F. J. (1969). Extensions of Lagrange multipliers in nonlinear programming. *SIAM J. appl. Math.* **17**, 1280–97.
Gould, F. J. & Tolle, J. W. (1971). A necessary and sufficient qualification for constrained optimization. *SIAM J. appl. Math.* **20**, 164–72.

Gould, F. J. & Tolle, J. W. (1975). Optimality conditions and constraint qualifications in Banach space. *JOTA* **15**, 667–84.

Hadley, G. (1964). *Nonlinear and dynamic programming*. Addison-Wesley, Reading, Mass.

Halkin, H. (1974). Necessary conditions in mathematical programming and optimal control theory. In Kirby, B. J. (ed.), *Optimal control theory and its applications*, part I. Lecture Notes in Economics and Mathematical Systems 105, pp. 113–65. Springer, Berlin.

Hestenes, M. R. (1966). *Calculus of variations and optimal control theory*. Wiley, New York.

Hoàng Tuy (1974*a*). On the convex approximation of nonlinear inequalities. *Math. Operationsforsch. Statist.* **5**, 451–66.

Hoàng Tuy (1974*b*). On a general minimax theorem. *Sov. Math. Dokl.* **15**, 1689–93 (English translation from the Russian).

Holmes, R. B. (1972). *A course on optimization and best approximation*. Lecture Notes in Mathematics 257. Springer, Berlin.

Ioffe, A. D. & Tikhomirov, V. M. (1968). Duality of convex functions and extremum problems. *Russian Math. Surveys* **23**, 53–124 (English translation).

Jongen, H. Th. (1977). *On non-convex optimization* (thesis). Technological University of Twente.

Kakutani, S. (1941). A generalization of Brouwer's fixed point theorem. *Duke Math. J.* **8**, 457–9.

Kall, P. (1976). *Stochastic linear programming*. Springer, Berlin.

Karamardian, S. (1969). The nonlinear complementarity problem with applications I, II. *JOTA* **4**, 87–98; 167–81.

Karamardian, S. (1971). Generalized complementarity problem. *JOTA* **8**, 161–8.

Karamardian, S. (1972). The complementarity problem. *Mathematical programming* **2**, 107–29.

Karamardian, S. (1976). An existence theorem for the complementarity problem. *JOTA* **19**, 227–32.

Karush, W. (1939). *Minima of functions of several variables with inequalities as side conditions* (master thesis). University of Chicago.

Kaulgud, N. N. & Pai, D. V. (1975). Fixed point theorems for set-valued mappings. *Nieuw Archief voor Wiskunde* **23**, 49–66.

Klee, V. L. (1948). The support property of a convex set in a normed linear space. *Duke Math. J.* **15**, 767–72.

Knaster, B., Kuratowski, C. & Mazurkiewicz, S. (1929). Ein Beweis des Fixpunktsatzes für *n*-dimensionale Simplexe. *Fund. Math.* **14**, 132–8.

Kuhn, H. W. (1960). Some combinatorial lemmas in topology. *IBM J. Res. Dev.* **4**, 518–24.

Ky Fan (1952). Fixed-point and minimax theorems in locally convex topological linear spaces. *Proc. natn. Acad. Sci. USA* **38**, 121–6.

Levine, P. & Pomerol, J.-Ch. (1974). Infinite programming and duality in topological vector spaces. *J. Math. Anal. Applics.* **46**, 75–89.

Liusternik, L. & Sobolev, V. (1961). *Elements of functional analysis*. Ungar, New York.

Luenberger, D. G. (1969). *Optimization by vector space methods*. Wiley, New York.

Luenberger, D. G. (1973). *Introduction to linear and nonlinear programming*. Addison-Wesley, Reading, Mass.

Magnanti, T. L. (1974). *Fenchel and Lagrange duality are equivalent.* Mathematical Programming 7, 253–8.

Mancino, O. G. & Stampacchia, G. (1972). Convex programming and variational inequalities. *JOTA* **9**, 3–23.

Mangasarian, O. L. (1969). *Nonlinear programming.* McGraw-Hill, New York.

Mangasarian, O. L. (1974). *Equivalence of the complementarity problem to a system of nonlinear equations.* Comp. Sci. Rep. 227, University of Madison, Wisconsin.

Mangasarian, O. L. (1975*a*). *Solution of linear complementarity problems by linear programming.* Comp. Sci. Rep. 257, University of Madison, Wisconsin.

Mangasarian, O. L. (1975*b*). Unconstrained Lagrangians in nonlinear programming. *SIAM J. Control* **13**, 772–91.

Mangasarian, O. L. & Ponstein, J. (1965). Minmax and duality in nonlinear programming. *J. Math. Anal. Applics.* **11**, 504–18.

McCormick, G. P. (1967). Second order conditions for constrained minima. *SIAM J. appl. Math.* **15**, 641–52.

McLinden, L. (1974). An extension of Fenchel's duality theorem to saddle functions and dual minimax problems. *Pacific J. Math.* **50**, 135–58.

McLinden, L. (1975). Duality theorems and theorems of the alternative. *Proc. Am. Math. Soc.* **53**, 172–5.

Motzkin, T. S. (1936). *Beiträge zur Theorie der linearen Ungleichungen* (inaugural dissertation). Azriel, Jerusalem.

Neustadt, L. W. (1974). *Optimization.* Princeton University Press.

Nieuwenhuis, J. W. (1978). *Duality results in mathematical programming* (thesis). University of Groningen.

Pallu de la Barrière, R. (1967). *Optimal control theory.* Saunders, Philadelphia, Penn.

Peterson, E. L. (1973). Geometric programming and some of its extensions. In Avriel, M., Rijckaert, M. J. & Wilde, D. J. (eds.) *Optimization and design*, pp. 228–89. Prentice-Hall, Englewood Cliffs, N.J.

Pierre, D. A. & Lowe, M. J. (1975). *Mathematical programming via augmented Lagrangians, an introduction with computer programs.* Addison-Wesley, Reading, Mass.

Ponstein, J. (1966). Existence of equilibrium points in nonproduct spaces. *SIAM J. appl. Math.* **14**, 181–90.

Ponstein, J. (1967). Seven kinds of convexity. *SIAM Rev.* **9**, 115–19.

Ponstein, J. (1968). Multiplier functions in optimal control. *SIAM J. Control* **6**, 648–58.

Ponstein, J. & Klein Haneveld, W. K. (1975). On a general saddle-point condition in normed spaces. *Mathematical Programming* **9**, 118–22.

Pontryagin, L. S., Boltyanskii, V. G., Gramkelidze, R. V. & Mishchenko, E. F. (1962). *The mathematical theory of optimal processes.* Wiley, New York.

Rockafellar, R. T. (1970*a*). Conjugate convex functions in optimal control and the calculus of variations. *J. Math. Anal. Applics.* **32**, 174–222.

Rockafellar, R. T. (1970*b*). *Convex Analysis.* Princeton University Press.

Rockafellar, R. T. (1974*a*). Augmented Lagrange multiplier functions and duality in nonconvex programming. *SIAM J. Control.* **12**, 268–85.

Rockafellar, R. T. (1974*b*). *Conjugate duality and optimization.* Regional conference series in applied mathematics 16. SIAM, Philadelphia, Penn.

Rockafellar, R. T. (1976*a*). Augmented Lagrangians and applications of the proximal point algorithm in convex programming. *Math. O.R.* **1**, 97–116.

Rockafellar, R. T. (1976*b*). Monotone operators and the proximal point algorithm. *SIAM J. Control Opt.* **14**, 877–98.

Rockafellar, R. T. & Wets, R. J.-B. (1976*a*). Stochastic convex programming: basic duality. *Pacific J. Math.* **62**, 173–95.

Rockafellar, R. T. & Wets, R. J.-B. (1976*b*). Stochastic convex programming: singular multipliers and duality. *Pacific J. Math.* **62**, 507–22.

Ritter, K. (1969, 1969, 1970). Optimization theory in linear spaces I, II, III. *Math. Ann.* **182**, 189–206; **183**, 169–80; **184**, 133–54.

Ritter, K. (1967). Duality for nonlinear programming in a Banach space. *SIAM J. appl. Math.* **5**, 294–302.

Robertson, A. P. & Robertson, W. J. (1973). *Topological vector spaces.* Cambridge University Press.

Roode, J. D. (1968). *Generalized Lagrangian functions in mathematical programming* (thesis). University of Leiden.

Russell, D. L. (1970). *Optimization theory.* Benjamin. New York.

Saigal, R. (1976). Extension of the generalized complementarity problem. *Math. O.R.* **1**, 260–6.

Schaafsma, W. (1970). Most stringent and maximin tests as solutions of linear programming problems. *Z. Wahrscheinlichkeitstheorie verw. Geb.* **14**, 290–307.

Sion, M. (1958). On general minimax theorems. *Pacific J. Math.* **8**, 171–6.

Slater, M. (1951*a*). *Lagrange multipliers revisited: a contribution to nonlinear programming.* Cowles Commission Discussion Paper Mathematics 403, Yale University. Also as RAND-report RM 676, The RAND Corporation, Santa Monica, Cal.

Slater, M. (1951*b*). A note on Motzkin's transposition theorem. *Econometrica* **19**, 185–6.

Smart, D. R. (1974). *Fixed point theorems.* Cambridge University Press.

Smith, R. H. & Vandelinde, D. (1972). A saddle-point optimality criterion for nonconvex programming in normed spaces. *SIAM J. appl. Math.* **23**, 203–13.

Stiemke, E. (1915). Uber positive Lösungen homogener linearer Gleichungen. *Math. Ann.* **76**, 340–2.

Stoer, J. & Witzgall, C. (1970). *Convexity and optimization in finite dimensions I.* Springer, Berlin.

Tabak, D. & Kuo, B. C. (1971). *Optimal control by mathematical programming.* Prentice-Hall, Englewood Cliffs, N.J.

Taylor, P. D. (1973). Subgradients of a convex function obtained from a directional derivative. *Pacific J. Math.* **44**, 739–47.

Tucker, A. W. (1956). Dual systems of homogeneous linear relations. In Kuhn, H. W. & Tucker, A. W. (eds.) *Linear inequalities and related systems.* Annals of Mathematics Studies 38, pp. 3–18. Princeton University Press.

Vajda, S. (1961). *Mathematical programming.* Addison-Wesley, Reading, Mass.

Van Slyke, R. M. & Wets, R. J.-B. (1968). A duality theory for abstract mathematical programs with applications to optimal control theory. *J. Math. Anal. Applics.* **22**, 679–706.

Varaiya, P. P. (1967). Nonlinear programming in Banach space. *SIAM J. appl. Math.* **15**, 284–93.
Wagner, H. M. (1969). *Principles of operations research.* Prentice-Hall, Englewood Cliffs, N.J.
Wets, R. (1976). *Grundlagen konvexer Optimierung.* Lecture Notes in Economics and Mathematical Systems 137. Springer, Berlin.

Index